NUCLEOSYNTHESIS

NUCLEOSYNTHESIS

Proceedings of a conference held January 25–26, 1965,

at the Institute for Space Studies, Goddard Space Flight Center,

NASA, New York

Edited by

W. David Arnett, Carl J. Hansen, J. W. Truran,

and A. G. W. Cameron

GORDON AND BREACH SCIENCE PUBLISHERS

NEW YORK · LONDON · PARIS

150 Fifth Avenue, New York, N. Y. 10011

Library of Congress Catalog Card Number 67–28 241

Editorial Office for Great Britain:
Gordon and Breach Science Publishers Ltd., 8 Bloomsbury Way, London WC1, England

Editorial Office for France:
Gordon & Breach
7-9 rue Emile Dubois, Paris 14^e^

Distributed in France by:
Dunod Editeur, 92, rue Bonaparte, Paris 6^e^

Distributed in Canada by:
The Ryerson Press, 299 Queen Street West, Toronto 2B, Ontario

Contents

Introduction

On January 25–26, 1965, the Institute for Space Studies of the Goddard Space Flight Center, National Aeronautics and Space Administration, was host to an international group of astronomers and physicists gathered to discuss the problems of nucleosynthesis. This was the seventh in a continuing series of interdisciplinary meetings on topics in space physics held at the Institute. The conference was organized by A. G. W. Cameron of the Goddard Institute for Space Studies.

The proceedings of this conference were tape-recorded and a transcript was made. From this transcript, first drafts of the papers were prepared in order to assist the authors in readying their papers for publication. We wish to thank the authors for their cooperation. We are most grateful to Mrs. E. Silva, the conference secretary, who handled very capably the arrangements for the conference and subsequently made important contributions to the preparation and typing of the manuscripts. Many thanks are also due to Miss A. Goldshein, who prepared and typed the final manuscripts. Mr. G. Goodstadt, Mr. N. Panagakos, and Mr. L. J. Stein, and several other staff members of the Goddard Institute for Space Studies gave valuable assistance in the organization of the conference, and we wish to thank them for their help in making it a success.

HISTORICAL BACKGROUND

Theories of the origin of the elements are guided, in large measure, by our knowledge of the abundances of the elements. It is clear, therefore, that nucleogenesis is a young subject. The occurrence of natural radioactive elements was recognized early in this century. The presence of appreciable amounts of the radioactive parents implies that these elements were born at some point in time. A knowledge both of the abundances of the parent elements and their decay products and of the decay life-times allows us to determine the age of these elements. Estimates of the age range from five billion to fifteen billion years, depending on whether the parent nuclei are assumed to have been formed at a given time or, rather, to have been formed continuously.

Early studies of element abundances were limited, necessarily, to the earth. The development of spectroscopy in the 19th century provided a vast new area of investigation. The same spectral lines that were seen in laboratory studies could be found as absorption features in the solar spectrum. It was concluded that the elements of which the sun is composed are the same as those existing on the earth. Studies of the spectra of other stars led to similar conclusions. Although the presence of these elements was established, it was not possible to make accurate determinations of their relative abundances. It was therefore quite reasonable to assume that the composition of the galaxy is everywhere uniform.

Early theories of nucleogenesis

A universe of uniform chemical composition strongly suggests a single event theory of the origin of the elements. This led physicists to search within the framework of cosmology for a set of physical conditions which would produce the present abundance distribution. Three different theories were developed, the details of which have been summarized in a review article by Alpher and Herman (1950). As certain aspects of those early theories have survived, it is instructive to consider their essential features.

Most of the early treatments of the problem of element formation were governed by considerations of equilibrium. It was assumed that at an early stage in the universe the matter was gathered together in a high temperature, high density state. Under these conditions nuclear reactions take place very rapidly. The relative abundances of the various nuclear species for a specified temperature and density can be determined from statistical mechanics, provided we know the energies and spins of the ground and excited states of the nuclei involved. A major difficulty with these theories arises from the fact that for a specified temperature and density the equilibrium abundances are sharply peaked in a narrow range of mass number. For example, for temperatures of a few billion degrees and densities of the order of 10^6 gm cm^{-3}, the nuclear statistical equilibrium equations predict a peak in the iron region, which is consistent with the observed abundance distributions. However, under the same conditions the predicted abundances of the heavy elements fall many orders of magnitude below the observed values. Many attempts were made to construct a massive "stellar" model possessing the correct distribution of density and temperature so that the total composition of the body would have the observed composition of the elements. These models encountered the further difficulty that it was not possible to cool such massive bodies rapidly enough to "freeze in" the abundances of the elements with the composition characteristic of the original high densities and temperatures. Current theories of nucleosynthesis attribute only the iron peak to an equilibrium process, as will be seen in our subsequent discussion.

Mayer and Teller (1949) proposed a polyneutron fission theory of nucleogenesis that assumed the universe in its early state to consist of large bodies composed of a cold nuclear fluid. The surfaces of these bodies would be unstable and polyneutron fragments of various sizes would be cast loose. Nuclear processes proceeding on these fragments—fission, beta decay and neutron evaporation—would result in the production of heavy elements. It was found that the relative isotopic compositions of the heavy elements could be rather well reproduced, subject to the right choice of the parameters of the theory. The major problem with this theory is the formation of the polyneutron bodies. It is not clear how such bodies can arise in any cosmology or be formed in any stellar interior.

Alpher, Bethe, and Gamow (1948) proposed another non-equilibrium theory of nucleogenesis. This model pictured the early stage of matter as a compressed neutron gas. The expansion of this neutron fluid following an initial explosion would result in the decay of some of the neutrons into protons and electrons, the relative numbers of protons and neutrons being fixed at an early stage by the equilibrium between direct and inverse beta (electron) reactions. Subsequent neutron captures would then build up all of the heavier elements, with some readjustment due to beta decay.

In a continuation of this work, Alpher observed that the abundance peaks in the heavy element region corresponded to regions in which the neutron capture probabilities had been determined by Hughes (1946) to be small. This correlation made apparent the need for a neutron capture process in the synthesis of the heavy elements. A rapid neutron capture process alone cannot, however, account for the detailed features of the relative abundances of heavy isotopes. Furthermore, there are no stable nuclei with mass numbers five and eight, suggesting that the production of nuclei with mass numbers higher than helium ($A = 4$) cannot result from neutron capture alone.

Evidence for element synthesis in stars

These early theories assumed that the period of element formation was of brief duration, consistent with the belief held at that time that the observed element distribution was uniform throughhout the galaxy. While the need for nuclear processes in stars had been established by the calculations of Bethe and Critchfield (1938) of the reaction rates for the proton-proton and of Bethe (1939) and von Weizsäcker (1938) for the carbon-nitrogen cycle, it was not recognized that subsequent stages of nuclear burning in stellar interiors might contribute to the formation of the heavy elements observed in nature.

The recognition that nucleosynthesis was a continuing process in stellar interiors followed the discovery by Merrill (1952) of the presence of the element technetium in the atmospheres of red giant stars. As technetium has no stable isotopes (the longest lived isotope has a half-life of less than two million years), its presence in abundances sufficient to be observed suggested that element synthesis had taken place quite recently in those stars.

This conclusion was confirmed by the discovery in the mid–1950's, that there are general abundance differences between certain broad classes of stars, in the sense that the ratio of the abundances of the heavy elements to hydrogen was variable and this ratio could be correlated with the age of the star. It was found that the ratio of all elements with proton number $Z > 2$ to hydrogen was an increasing function of the time of formation of the star in the galaxy.

The presence in stars of anomalous abundances of elements which we believe are largely produced by a specific nuclear burning process provides somewhat more direct evidence for nucleosynthesis in stars. The presence of technetium in red giant stars, for example, strongly suggests that element synthesis by neutron capture has occurred recently in those stars. The high abundance of carbon in "Carbon stars" is believed to result from helium burning in the interior, assuming that mass has been carried to the surface by convection. The observational evidence available for the various modes of element synthesis is discussed by G. Wallerstein in these proceedings.

These observations suggest a simple model of galactic evolution. Assuming the primordial gas to be composed of hydrogen, with perhaps some small amount of helium, the first generation of stars will be formed with this composition. The evolution of these stars is characterized both by element synthesis during various stages of nuclear burning and by mass loss. In this manner, the heavy element content of the interstellar medium will be increased. Subsequent generations of stars will be formed from gas enriched in these heavy elements.

Mechanisms of mass loss

The enrichment of the interstellar medium in the products of the various nuclear burning stages of stellar evolution implies mass loss by these stars. There are three principal ways by which this mass loss is known to take place. The most spectacular mechanism is the supernova explosion. In such an explosion a major portion of the mass of the star is thought to be ejected, leaving behind an imploded remnant which, if stable, might become a neutron star. The detailed features of these explosions are currently under investigation (see the papers by D. Arnett and by S. A. Colgate in these proceedings). The rate of supernova explosions in an average galaxy is estimated to be approximately one event every 50–300 years.

A second mechanism of mass loss is the nova explosion. These are much less spectacular events than are supernovas, only about one part in 1000 of the stellar mass being cast off. However, many of the nova explosions are found to be recurrent in which case, during the course of many explosions, a substantial fraction of the stellar mass may be returned to the interstellar medium.

There is also evidence that mass loss takes place continuously during many stages of evolution. It is felt that mass loss can take place during the final stages of contraction of the star from the interstellar medium. A significant amount of mass loss has been observed for stars in the red giant phase. This continuous ejection of mass appears to be associated with hot stellar coronas and with magnetic fields on the stellar surfaces.

THE ABUNDANCES OF THE ELEMENTS

We wish to consider in detail the role of nuclear reactions in stars in the synthesis of the elements. A great deal of insight into these modes of element synthesis can be gained from a study of the observed abundances.

Early estimates of the abundances of the elements were based to a great extent on chemical analyses of various constituents of the earth's crust. This is clearly a poor choice of such information, due both to the fact that the earth has lost most of its volatile constituents and to the fact that the remaining material has been subjected to extreme chemical differentiation. It was evident, nevertheless, that the general features of the observed distribution of element abundances cannot be correlated with their chemical properties.

The best sources of information regarding the abundances of non-volatile elements are the meteorites. It is generally assumed that these bodies have not undergone the same degree of differentiation of their chemical constituents as has the earth. The degree of material differentiation is sufficient, however, to serve as a means of classification, viz; iron meteorites (siderites), stony-irons, and chondritic meteorites. Of these, the chondritic meteorites are thought to represent best the non-volatile constituents of the solar system.

The classic compilation of element abundances by Suess and Urey (1956) has played a major role in recent analyses of the modes of element synthesis. They relied heavily on studies of abundances in chondritic meteorites for the non-volatile constituents. The abundances of the volatile elements, notably H, He, C, N, O, and Ne, can best be determined from spectroscopic analyses of the light coming from stars and from gaseous nebulae. These analyses are complicated by the need both for accurate determinations of atomic oscillator strengths and for a detailed model for the stellar atmospheres. For both classes of elements, accurate relative isotopic abundances are available from mass spectroscopic data.

At the time that Suess and Urey made their abundance compilation, there were relatively few reliable determinations of abundances in chondritic meteorites. Some interpolation was required in order to construct a complete abundance pattern. Their primary criterion was that the abundances of nuclei of odd mass number should show a smooth variation with mass number. This procedure was helped by the fact that many elements contain two isotopes with odd mass numbers, establishing the slope of the abundance curve in these regions.

A revised compilation of element abundances has been presented by Cameron (1963). This compilation is based almost entirely on abundances measured in chondritic meteorites and, where available, in carbonaceous chondritic meteorites. The carbonaceous chondrites are a class of chondritic meteorites which have been subjected to a minimum of chemical differentiation. As much as half of their mass may be composed of water of crystallization and various complex carbon compounds. They also contain semi-volatile elements such as mercury which may be entirely missing in the ordinary chondrites. The abundances of the light volatile elements must still be determined from studies of the sun and stars. A comparison of the abundance compilation by Cameron with that by Suess and Urey is given in Table I. The abundances arrived at by Cameron are plotted as a function of mass number in Figure 1.

Table I

Comparison of Abundance Compilations

Element	*Suess-Urey*	*Cameron*
1 H	4.00×10^{10}	3.2×10^{10}
2 He	3.08×10^{9}	2.6×10^{9}*
3 Li	100	38
4 Be	20	7
5 B	24	6
6 C	3.5×10^{6}	1.66×10^{7}
7 N	6.6×10^{6}	3.0×10^{6}
8 O	2.15×10^{7}	$2.9 \quad 10^{7}$
9 F	1600	$\sim 10^{3}$*
10 Ne	8.6×10^{6}	2.9×10^{6}*
11 Na	4.38×10^{4}	4.18×10^{4}
12 Mg	9.12×10^{5}	1.046×10^{6}
13 Al	9.48×10^{4}	8.93×10^{4}
14 Si	1.00×10^{6}	1.00×10^{6}
15 P	1.00×10^{4}	9320
16 S	3.75×10^{5}	6.0×10^{5}
17 Cl	8850	1836
18 A	1.5×10^{5}	2.4×10^{5}
19 K	3160	2970
20 Ca	4.90×10^{4}	7.28×10^{4}
21 Sc	28	29
22 Ti	2440	3140
23 V	220	590
24 Cr	7800	1.20×10^{4}
25 Mn	6850	6320
26 Fe	6.00×10^{5}	8.42×10^{5}
27 Co	1800	2290
28 Ni	2.74×10^{4}	4.44×10^{4}
29 Cu	212	39
30 Zn	486	202
31 Ga	11.4	9.05
32 Ge	50.5	134
33 As	4.0	4.4
34 Se	67.6	18.8
35 Br	13.4	3.95
36 Kr	51.3	20
37 Rb	6.5	5.0
38 Sr	18.9	21
39 Y	8.9	3.6
40 Zr	54.5	23
41 Nb	1.00	0.81

Table I (*continued*)

Element	*Suess-Urey*	*Cameron*
42 Mo	2.42	2.42
44 Ru	1.49	1.58
45 Rh	0.214	0.26
46 Pd	0.675	1.00
47 Ag	0.26	0.26
48 Cd	0.89	0.89
49 In	0.11	0.11
50 Sn	1.33	1.33
51 Sb	0.246	0.15
52 Te	4.67	3.00
53 I	0.80	0.46
54 Xe	4.0	3.15
55 Cs	0.456	0.25
56 Ba	3.66	4.0
57 La	2.00	0.38
58 Ce	2.26	1.08
59 Pr	0.40	0.16
60 Nd	1.44	0.69
62 Sm	0.664	0.24
63 Eu	0.187	0.083
64 Gd	0.684	0.33
65 Tb	0.0956	0.054
66 Dy	0.556	0.33
67 Ho	0.118	0.076
68 Er	0.316	0.21
69 Tm	0.0318	0.032
70 Yb	0.220	0.18
71 Lu	0.050	0.031
72 Hf	0.438	0.16
73 Ta	0.065	0.021
74 W	0.49	0.11
75 Re	0.135	0.054
76 Os	1.00	0.73
77 Ir	0.821	0.500
78 Pt	1.625	1.157
79 Au	0.145	0.13
80 Hg	0.284	0.27
81 Tl	0.108	0.11
82 Pb	0.47	2.2
83 Bi	0.144	0.14
90 Th		0.069
92 U		0.042

* Revised abundances due to Cameron in 1966.

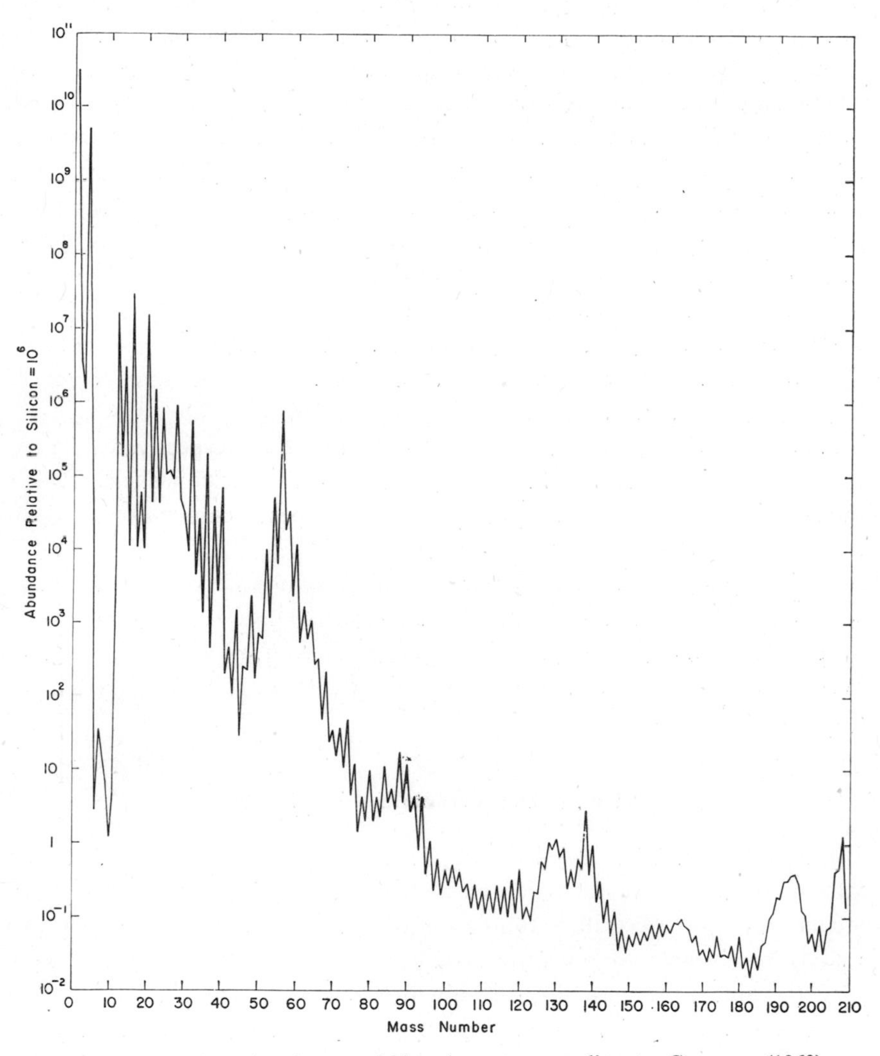

Fig. 1. The relative abundances of the elements according to Cameron (1963) are plotted as a function of mass number (the number of protons and neutrons in each nucleus).

The general features of these two compilations are the same. The most abundant element is hydrogen, followed closely by helium. The elements lithium, beryllium and boron are low in abundance but the products of helium burning, carbon and oxygen, are present in considerable amounts. The abundances in the mass region $24 \leqq A \leqq 32$ are roughly 10^{-2} to 10^{-1} of the abundances of C^{12} and O^{16}. The general decrease in the abundance level through mass number $A = 100$ is interrupted by a broad abundance peak centered on Fe^{56}. This iron peak is generally attributed to the equilibrium process. It should be noted that the abundance of Fe^{56} relative to its neighbors determined from meteoritic data is approximately a factor of five greater than the solar abundance (Aller, 1961).

Relative to the general decrease in abundances through mass number $A = 100$, the abundance level in the heavy element region is rather constant. Superimposed on this trend are peaks corresponding to nuclides with neutron numbers of 50, 82 and 126 ($A \sim 88$, 138 and 208, respectively). These neutron numbers are known to represent closed nucleon shells, which imply increased stability. We shall see how these peaks can be produced by neutron capture processes.

Pronounced deviations from the 'cosmic' abundance compilation presented above have been observed in stellar spectra. Discussions of the abundances in metallic line and peculiar A stars are given by W. L. W. Sargent, W. P. Bidelman, P. S. Conti, J. Faulkner, and H. L. Helfer in these proceedings.

NUCLEOSYNTHESIS IN STARS

Burbidge, Burbidge, Fowler and Hoyle (1957) discussed in detail the various nuclear processes which play a role in the synthesis of elements in stellar interiors. According to their model, charge-particle reactions are mainly responsible for element production through iron, beyond which neutron capture becomes the predominant mechanism. Eight distinct nuclear processes were required for the synthesis of the elements from hydrogen. Subsequent studies by these authors and by Cameron (1963) and others have modified various aspects of this model.

In general the history of a star is defined by a succession of stages of gravitational contraction and nuclear burning. During the contraction the interior of the star is heated by the release of gravitational potential energy. When the temperature is increased to the point at which the nuclear fuel present in the medium can begin to burn, the contraction is halted. Thermonuclear reactions then provide the source of energy generation necessary to maintain the stellar luminosity. With the exhaustion of this nuclear fuel, gravitational contraction resumes.

Detailed models of stellar evolution through the helium burning phase have been discussed by I. Iben in these proceedings. Further, R. Stothers has considered the evolution of a 30 solar mass star. H. Y. Chiu has discussed the final stages of presupernova evolution. For a more complete picture of our current understanding of stellar evolution, the reader is referred to the proceedings of an earlier conference held at this Institute (Stein and Cameron 1966).

The destruction of the light elements

During its initial contraction from the interstellar medium, the star is heated by the release of gravitational potential energy. When the temperature reaches 10^6 °K any deuterium present in the medium is destroyed by thermonuclear reactions with itself and with hydrogen. After the exhaustion of the deuterium, the star will continue to contract and its central temperature will continue to increase. In rapid succession the elements lithium, beryllium and boron will be destroyed by thermonuclear reactions with hydrogen.

Hydrogen burning

At slightly higher temperatures hydrogen thermonuclear reactions will be initiated. The contraction of the star is halted, and the star settles down to a long period of stability while the hydrogen in the interior is converted into helium. The major portion of the active lifetime of a star is spent in this hydrogen burning stage, defining the main sequence.

There are two basic sequences of reactions by which hydrogen can be converted into helium. Stars of mass $\leqq 2\ M_\odot$ ($M_\odot$ being the mass of the sun) burn hydrogen at temperatures $\leqq 1.7. \times 10^7$°K by the proton-proton chains. These reactions are summarized in Table II. The net result of this burning phase is the fusion of four protons to one He^4 nucleus, with the release of 26,730 million electron volts (MeV) of energy, or 6.68 MeV. per nucleon. In contrast, subsequent nuclear transformations by which the helium is converted to iron release only $\sim$2.22 MeV. per nucleon.

For stars more massive than $\sim 2\ M_\odot$, the initial contraction will continue until the central temperature exceeds 1.7×10^7 °K. At these temperatures, hydrogen burning by the carbon-nitrogen-oxygen cycles provides more energy generation than do the proton-proton chains. In these reactions the

Table II

Hydrogen burning reactions: the proton-proton chains

PP I	$H^1(p, \beta^+\nu)D^2$
	$D^2(p, \gamma)He^3$
	$He^3(He^3, 2p)He^4$
PP II	$He^3(\alpha, \gamma)Be^7$
	$Be^7(e^-, \nu)Li^7$
	$Li^7(p, \gamma)Be^8(2\alpha)$
PP III	$Be^7(p, \gamma)B^8$
	$B^8(\beta^+\nu)Be^8(2\alpha)$

carbon and nitrogen act as catalysts for the conversion of hydrogen to helium, as illustrated in Table III. Further, for stars of several solar masses, which burn their central hydrogen at temperatures of $\sim 3 \times 10^7$ °K, the isotopes of oxygen and F^{19} will be largely transformed into CNO-cycle nuclei.

Table III

Hydrogen burning reactions: the CNO cycle. (The arrows indicate a recycling of CNO nuclei)

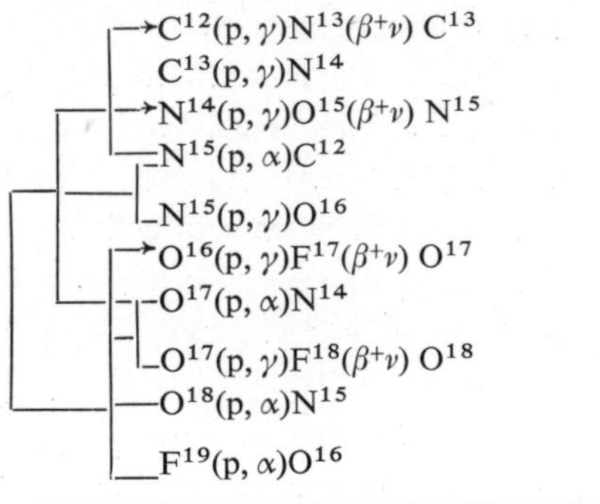

Calculations of the abundances of carbon, nitrogen and oxygen in equilibrium for the CNO cycle predict that N^{14} will be the most abundant nucleus. This is an important result with regard to our subsequent discussion of neutron capture, as N^{14} may provide an important source of neutrons for the synthesis of the heavy elements. These equilibrium calculations also predict a ratio $C^{12}/C^{13} \sim 4$ over a wide range of temperatures. There is good observational confirmation that the ratio C^{12}/C^{13} reaches values close to four in carbon stars, suggesting that this material has been processed by the CNO cycle (see the discussion by G. Wallerstein in these proceedings).

Helium burning

The exhaustion of the hydrogen fuel in the core is followed by the contraction of the core, the temperature increasing until the helium in the core ignites. Helium burning proceeds at temperatures $\geqq 10^8$ °K by means of the triple-alpha reaction

$$3\alpha \rightarrow C^{12}$$

The carbon formed in this way can capture another alpha-particle to form O^{16}

$$C^{12}(\alpha, \gamma)\, O^{16}$$

The production of heavier nuclei is impeded by the slow rate for alpha-particle capture by O^{16}.

Uncertainties associated with the rate of alpha capture by C^{12} make it difficult to determine with any accuracy the final abundances of the products of helium burning. The greater sensitivity of the rate of the triple-alpha reaction to the abundance of alpha-particles, however, suggests that if helium burning goes to completion O^{16} should be the major product.

Heavy ion thermonuclear reactions

Following helium burning, the core is composed predominantly of the nuclei C^{12} and O^{16}. The next nuclear burning stage must involve the interactions of these heavy ions with themselves.

The destruction of C^{12} will proceed at temperatures $\geqq 7 \times 10^8$ °K by the reaction $C^{12} + C^{12}$. The dominant reactions are

$$\begin{aligned} C^{12} + C^{12} &\rightarrow Na^{23} + p \\ &\rightarrow Ne^{20} + \alpha \end{aligned}$$

where the products are in roughly equal amounts. There is also a weak branching to $Mg^{23} + n$, but the direct effects of this neutron emission are of minor importance at these temperatures.

At slightly higher temperatures ($T \geqq 10^9$ °K) oxygen burning by $O^{16} + O^{16}$ is also possible. The dominant reactions in this case are

$$\begin{aligned} O^{16} + O^{16} &\rightarrow P^{31} + p \\ &\rightarrow Si^{28} + \alpha \\ &\rightarrow S^{31} + n \end{aligned}$$

Experimental data for these reactions is quite limited. A thorough discussion of these rates has been given by Reeves (1965; see also the article in these proceedings).

The equilibrium process

The protons and alpha-particles released in the heavy ion reactions will rapidly be recaptured. The principal final products of this nuclear burning stage should then be the alpha nuclei Mg^{24}, Si^{28} und S^{32}. Core contraction at this stage results in a further increase in the temperature.

If the temperature in the core can rise above 3×10^9 °K, then a great variety of nuclear reactions can occur with very rapid rates. The determination of these rates from the statistical properties of nuclei is discussed in the paper by Cameron et al. The products of the earlier burning stages, now present mainly in the form of the more stable Si^{28}, can be photodisintegrated at these temperatures. The protons, neutrons and alpha-particles released in this manner can then be captured on the remaining nuclei resulting in a buildup toward the iron region. This sequence of reactions has been considered in detail in the paper by Gilbert et al. in these proceedings.

The rapidity of the various reaction rates under these conditions suggests that the relative abundances of the various nuclear species should come into equilibrium. These equilibrium abundances can be calculated from statistical mechanics, as was done in the earlier theories. Figure 2 shows an equilibrium fit to the iron peak region by Cameron (1963) for a temperature of 4×10^9°K ($T_9 = 4$) and a density $\varrho = 10^6$ grams per cubic centimeter. The meteoritic abundances have been modified by reducing the abundance of iron relative to its neighbors by a factor of ~ 5 to agree with the solar value. These results are in reasonably good agreement in the immediate vicinity of iron. Detailed calculations of the equilibrium composition of matter at extremely high densities have been presented by S. Tsuruta in these proceedings.

Supernova explosions

The nuclear burning stages discussed in the previous paragraphs have been exoergic; that is, the net result of the various nuclear reactions has been

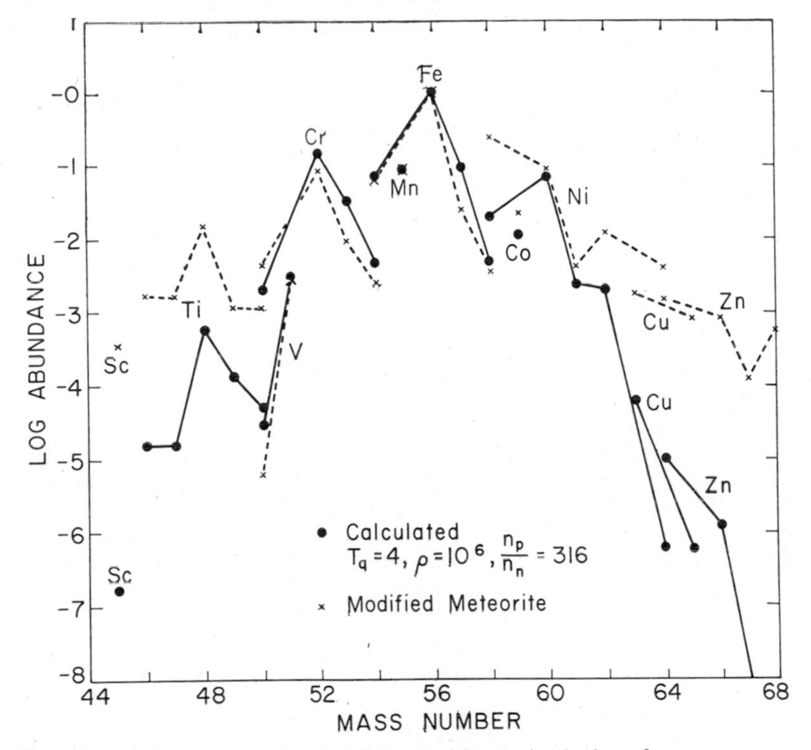

Fig. 2. An equilibrium calculation of the abundances in the iron for a temperature of 3×10^9 °K and a density $\varrho = 10^6$ gm/c.c. is compared to the observed abundances. Meteoritic abundances have been employed, except that the iron abundance has been reduced to agree with the solar value. The ratio of the numbers of free protons to free neutrons for this fit was $n_p/n_n = 316$.

the formation of heavier nuclei with the release of energy. This energy generation was sufficient to provide the stellar luminosity. However, Fe^{56}, the most abundant nucleus formed in the equilibrium process, has the largest binding energy per nucleon of any nucleus. Any further nuclear processing of this Fe^{56} will therefore result in a loss of energy by the star.

The only source of energy available is gravitational potential energy. As the core contracts, the temperature will increase. At temperatures above $\sim 6 \times 10^9$ °K, for a density of $\sim 10^6$ gm/c.c., the endoergic conversion of Fe^{56} to helium will proceed. This helium production must then be accompanied by a further contraction, or the collapse of the core.

There is another mechanism by which the collapse of the core might be initiated. At the high densities and temperatures existing in the core, electron capture by nuclei and free protons can proceed rapidly. The decrease in the electron number density will be accompanied by a decrease in the electron pressure which may be sufficient to trigger the collapse.

As the collapse proceeds, the density in the core can approach nuclear densities. Neutrinos released under these conditions can diffuse outward, depositing energy in the outer regions of the core. This deposition of energy can result in the formation of a shock wave in the region surrounding the core. If this shock is sufficiently energetic, the outer layers of the star can be blown off with high velocities in a supernova explosion. By this means, the interstellar medium may be enriched in the abundances of heavy elements. The characteristics of these shock waves are currently under investigation (see the papers by W. D. Arnett and by S. A. Colgate in these proceedings).

The passage of the shock wave through the outer regions of the star will result in an increase in both the temperature and the density of the gas. It has been proposed that a large amount of element synthesis might take place under these conditions. Recent calculations carried out by J. W. Truran indicate that regions composed initially of C^{12}, O^{16} or Si^{28} can be processed rapidly to nuclei in the iron peak. Neutron capture processes may also be important under these conditions.

Deuterium, lithium, beryllium and boron

We have seen that the elements deuterium, lithium, beryllium and boron are destroyed rapidly by interactions with protons during the initial contraction stage of a star. In general, these elements are known to be unstable at the temperatures characteristic of stellar interiors.

Several mechanisms have been proposed for the production of these light elements. Fowler, Greenstein and Hoyle (1962) have suggested that lithium, beryllium and boron might be formed by high energy spallation reactions during the early history of the solar system. Cameron has suggested that the passage of a supernova shock wave through a region of helium might result in the production of these elements. Further studies of this problem are required in order to decide whether either of these two mechanisms is correct or, rather, whether an entirely different mechanism is responsible for the production of the light elements.

Element synthesis by neutron capture

The buildup of the abundances of nuclei heavier than iron takes place predominantly by neutron capture. In general these neutron capture processes can be divided into two distinct groups, defined by the relative stability of the product nuclei against electron (beta) emission. If the products have beta decay half-lives longer than the half-lives for destruction by neutron capture, they can be regarded as stable in the neutron capture sense. This corresponds to the process of neutron capture on a fast time scale (the *r*-Process of Burbidge *et al.*). If the half-life for beta decay is shorter than the half-life for neutron destruction, then this decay can be assumed to take place before the capture of a neutron. This is neutron capture on a slow time scale (the *s*-process).

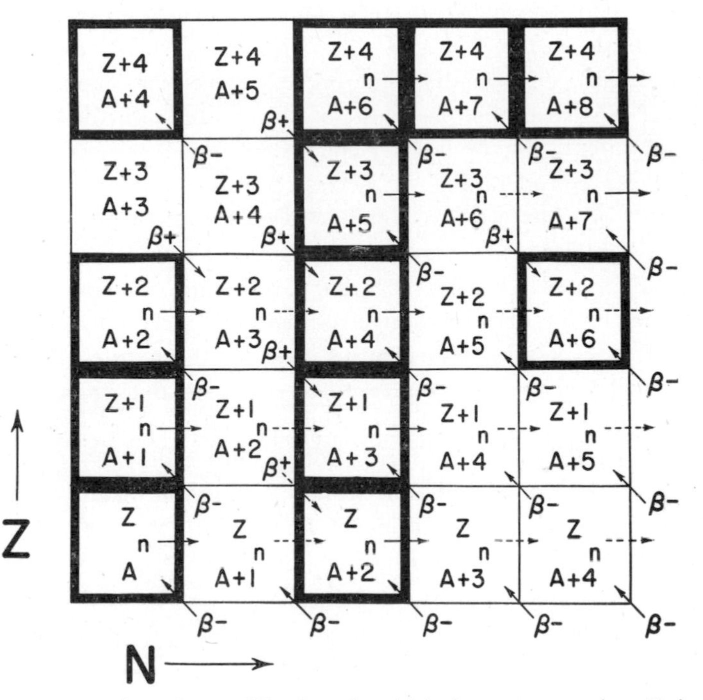

Fig. 3. A typical section of a nuclide chart in which the proton number, Z, is plotted vertically and the neutron number, N, horizontally.

A typical section of a nuclide chart is shown in Figure 3; beta stable nuclei are indicated by heavy black borders. In this figure the proton number Z is plotted vertically and the neutron number N is plotted horizontally. The mass number, the total number of protons and neutrons, is then given by $A = Z + N$. Neutron capture proceeds horizontally, along lines of constant Z. Beta decay processes, the decay of a neutron to a proton, involve no change in mass number and therefore proceed along the diagonals as

$$(Z, A) \rightarrow (Z + 1, A) + \beta^- + \bar{\nu}$$

where $\bar{\nu}$ is an antineutrino and β^- is an electron.

If we begin with the nucleus in the lower left hand corner (Z, A) and capture a neutron, the product nucleus is beta unstable. If the beta decay half-life is long compared to the mean time between captures, neutron capture will continue through the nuclei $(Z, A + 1)$ and $(Z, A + 2)$. If the neutron capture times are long, then the nucleus $(Z, A + 1)$ will beta decay to $(Z + 1, A + 1)$ which can then undergo neutron capture and subsequent beta decay to $(Z + 2, A + 2)$. This sequence of neutron captures and beta decays corresponds to the process of neutron capture on a slow time scale. It is clear that the path for this process will proceed through regions of stability for the heavy elements (the "valley of beta stability").

From these considerations, it is clear that we can determine a unique capture path provided we have some knowledge of the neutron capture and beta decay lifetimes. The beta decay rates are not extremely sensitive to the physical conditions. On the other hand, the rate of neutron capture is proportional to the number of neutrons available per unit volume in the medium. It is necessary, therefore, to consider the nature of the neutron sources available for neutron capture. Accurate determinations of the neutron capture cross sections are also essential (see the papers by J. H. Gibbons, G. I. Bell and P. A. Seeger in these proceedings).

Neutron capture on a slow time scale is believed to take place at various presupernova stages of evolution (D. D. Clayton, these proceedings). In the early stages of helium burning, $T_8 \sim 1$, the $C^{13}(\alpha, n)O^{16}$ reaction provides a source of neutrons. At somewhat higher temperatures, a significant flux of neutrons can result the reaction $Ne^{22}(\alpha, n)Mg^{25}$. This Ne^{22} is thought to be formed by the following sequence of reactions

$$N^{14}(\alpha, \gamma)\, F^{18}(\beta^+\nu)O^{18}$$

$$O^{18}(\alpha, \gamma)Ne^{22}$$

taking place during helium burning, where the N^{14} represents the principal product of hydrogen burning by the CNO-cycles. Further neutron capture processing can take place during carbon burning, where the protons and alpha-particles released in the $C^{12} + C^{12}$ reaction can initiate the following sequence of reactions

$$C^{12}(p, \gamma)N^{13}(\beta^+\nu)\, C^{13}(\alpha, n)O^{16}$$

This source of neutrons is less certain as, for temperatures greater than 7.5×10^8 °K, the N^{13} will be destroyed by $N^{13}(\gamma, p)C^{12}$ before it can β^+ decay.

Neutron capture on a fast time scale is thought to take place in supernova explosions. It is felt that large fluxes of neutrons may be produced under these extreme conditions. Current studies of this problem by J. W. Truran indicate that the extent of element synthesis by neutron capture under these conditions is rather sensitive to the detailed characteristics of the supernova explosion, which are still rather poorly known. Calculations of the *r*-process abundances have been carried out by Seeger et al (1965, see also the paper by P. A. Seeger in these proceedings).

The abundances in the heavy element region are displayed in Figure 1. Superimposed on this abundance pattern are peaks at $A \sim 88$, 138 and 208 corresponding to nuclei with neutron numbers 50, 82 and 126. These peaks can be accounted for by neutron capture on a slow time scale. Neutron numbers 50, 82 and 126 correspond to stable, closed shell nucleon configurations. The neutron capture probabilities for these nuclei are small. Nuclei in these regions, in the absence of a large neutron flux, will tend to beta decay before they can capture another neutron. Thus there is a tendency to accumulate nuclei at the closed shell positions.

In Figure 1 we note the presence of somewhat broader and lower subsidary peaks occurring about ten mass units before the closed shell peaks These features are attributed to neutron capture on a fast time scale. This neutron capture process will generally form nuclei in neutron rich regions far off the valley of beta stability. In these regions as well, the neutron capture probabilities will decrease at the closed shell positions. As the beta decay lifetimes may be comparable to the general time scale of the fast neutron capture process, there will be an accumulation of nuclei in the region immediately preceding the closed neutron shell positions.

The peaks at mass number $A \sim 103$ and in the rare earth region, $A =$ 150–170, are not explained by these mechanisms. Cameron (1963) has suggested that these peaks might result from fast neutron capture for a smaller value of the neutron flux. The mass peak at $A \sim 103$ has also been interpreted as resulting from the fission of the transuranic elements. There are also a number of heavy nuclei which cannot be formed by any of these neutron capture processes. These "by-passed" nuclei are thought to be produced by (p, γ) and (γ, n) reactions occurring perhaps, in supernova explosions.

REFERENCES

Aller, L. H. (1961). *The Abundances of the Elements.* Interscience Publishers, Inc., New York.

Alpher, R. A., Bethe, H. A. and Gamow, G. (1948). *Phys. Rev.* **73**, 803.

Alpher, R. A. and Herman, R. C. (1950). *Rev. Mod. Phys.* **22**, 153.

Bethe, H. A. and Critchfield, C. L. (1938), *Phys. Rev.* **54**, 248.

Bethe, H. A. (1939), *Phys. Rev.* **55**, 103.

Burbidge, E. M., Burbidge, G. R., Fowler, W. A. and Hoyle, F. (1957). *Rev. Mod. Phys.* **29**, 547.

Cameron, A. G. W. (1963), *Nuclear Astrophysics*, compilation of notes from lectures given at Yale University.

Fowler, W. A., Greenstein, J. L. and Hoyle, F. (1962). *Geophys. J. Roy. Astron. Soc.* **6**, 148.

Hughes, D. J. (1946). *Phys. Rev.* **70**, 106A.

Mayer, M. G. and Teller, E. (1949). *Phys. Rev.* **76**, 1226.

Merrill, P. W. (1952), *Science* **115**, 484.

Reeves, H. (1965). *Stellar Energy Sources*, Chap. 2 in Stars and Stellar Systems, Vol. VIII, Aller and McLaughlin eds., Un. of Chicago Press.

Seeger, P. A., Fowler, W. A. and Clayton, D. D. (1965). *Astrophys. J.* Suppl. **11**, 121.

Stein, R. F. and Cameron, A. G. W. (Eds). (1966), *Stellar Evolution*, Plenum Press.

Suess, H. E. and Urey, H. C. (1956), *Rev. Mod. Phys.* **28**, 53.

von Weizsäcker, C. F. (1938). *Physik Z.* **39**, 633.

Difficulties with Theories of Nucleosynthesis in Stars

HANS E. SUESS

The synthesis of the heavy elements, $A > 40$, in stellar interiors is generally attributed to four distinct nuclear processes (Burbidge, G. R., Burbidge, W. A., Fowler, and Hoyle, 1957, quoted below B^2HF; see also Cameron, 1963): (1) *s*-processes, producing the nuclides that are formed from a stable or almost stable nuclear species by neutron capture on a time scale long compared to beta decay half-lives; (2) *r*-processes, producing neutron-rich nuclides by rapid addition of neutrons on a time scale of the order of a second; (3) *e*-processes, yielding nuclides in relative amounts close to equilibrium concentrations, assuming a fixed temperature, density, and proton-to-neutron ratio; (4) *p*-processes, leading to the formation of neutron-poor nuclides by either (γ, n) or (p, γ) reactions proceeding on nuclides formed by neutron capture. It is generally assumed that the abundances of these heavy elements observed in nature, derived from analyses of meteorites and from spectral analyses of the sun, comprise a mixture of the products of these four nuclear processes (B^2FH, 1957).

Several observations, however, and particularly those of certain details in the abundance distribution, indicate that the origin of the solar system elements as a mixture of products from four different nuclear processes, is highly improbable. It should be emphasized, however, that this critique is valid only for the solar system elements for which abundances and accurate isotopic compositions are sufficiently well known so that these details can be recognized.

Since publication of the abundance table by Suess and Urey (1956), much new work has been done on trace element analyses of meteorites, and our knowledge of the chemical composition of meteorites has improved considerably. A careful and detailed analysis of the available analytical data is now being carried out by Urey and the author in order to compile a revised table of nuclear abundances. This preliminary study has confirmed the existence of regularities which cannot be understood in terms of a mixture of the products of the four nuclear processes defined previously (Suess, 1964).

I would like to indicate the kind of evidence from which one might conclude that the elements of our solar system have not resulted from a mixture of these four nuclear components.

THE e-PROCESS

The general features of the abundance distribution are illustrated in Figure 1. Superimposed on a general decrease in abundance with mass number are peaks at the neutron closed-shell positions, $N = 28$, $N = 50$, and $N = 82$. One important characteristic of these four peaks is a similarity in the character of the abundance distribution around the various shell

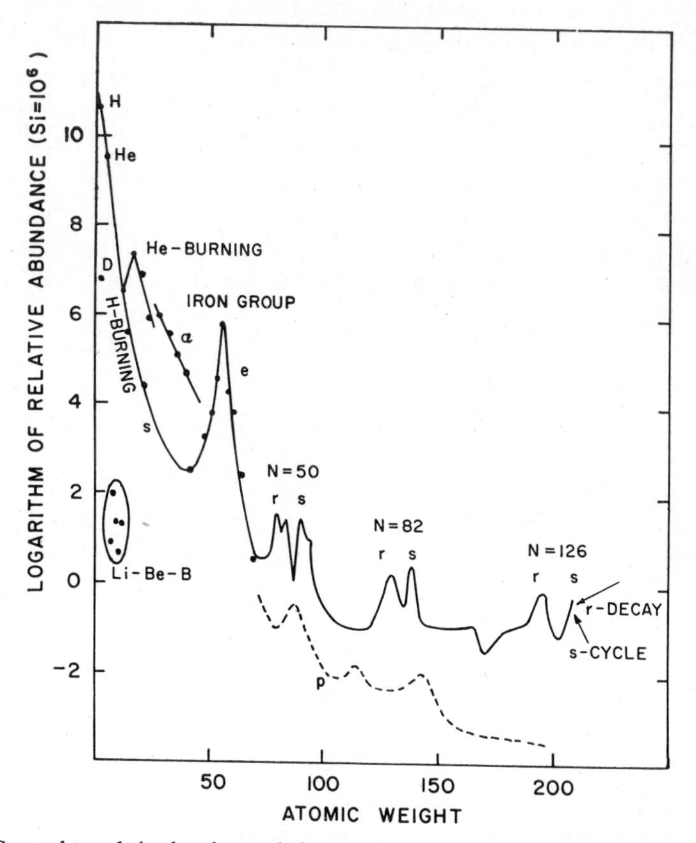

Fig. 1. General trends in the observed element abundances with mass numbers, according to B²HF. Details are omitted and, therefore, some intrinsic regularities cannot be recognized.

edges that can be taken as an indication speaking in favor of a single event theory for the origin of the elements: elements immediately preceeding a magic number of neutrons show normal isotopic composition, viz., a medium heavy isotope, adjacent to a stable odd A isotope, or a heavy isotope, possessing the highest abundance; elements following in atomic number a neutron shell closure always show an abnormal isotopic composition, and the lightest isotope has the highest abundance, i.e. Ni^{58}, Zr^{90} and Nd^{142}. These features are evident in the detailed plot of the element abundances (Suess and Urey, 1956) shown in Figure 2, but cannot be recognized in Figure 1.

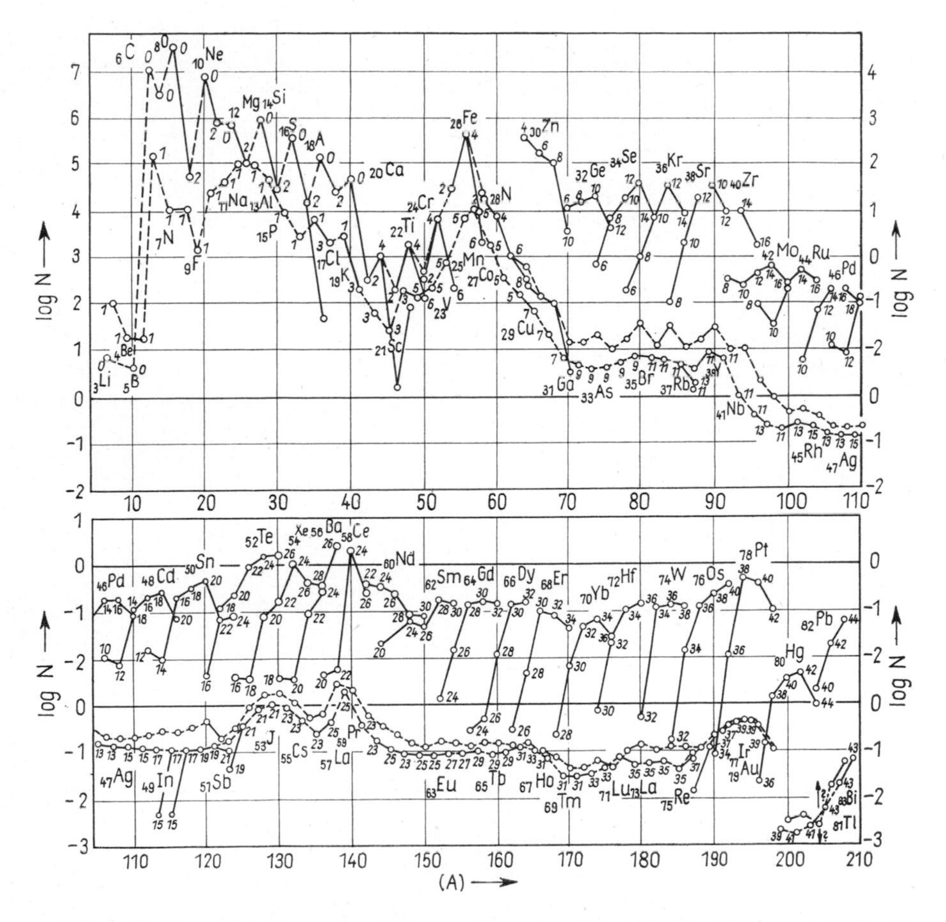

Fig. 2. Nuclear abundances according to Suess and Urey (1956). All the values for the individual nuclear species are shown and, therefore, the impression is confusing.

According to B^2FH (1957), the iron peak elements, in the neighborhood of $N = 28$, were formed by *e*-processes, whereas the relatively high abundances of nuclides containing 50 and 82 neutrons are assumed to have resulted from slow neutron buildup. If this picture were correct, then the similarities between the abundance distributions at the closed-shell positions described above would have to be assumed to be only coincidental.

THE s-PROCESS

There is evidence, as well, to suggest that the distinction between *s*-process and *p*-process elements may not be valid. In general, the abundances of nuclear species with the same neutron excess number ($I = N - Z$) show, within certain mass ranges, a smooth dependence with mass number. This

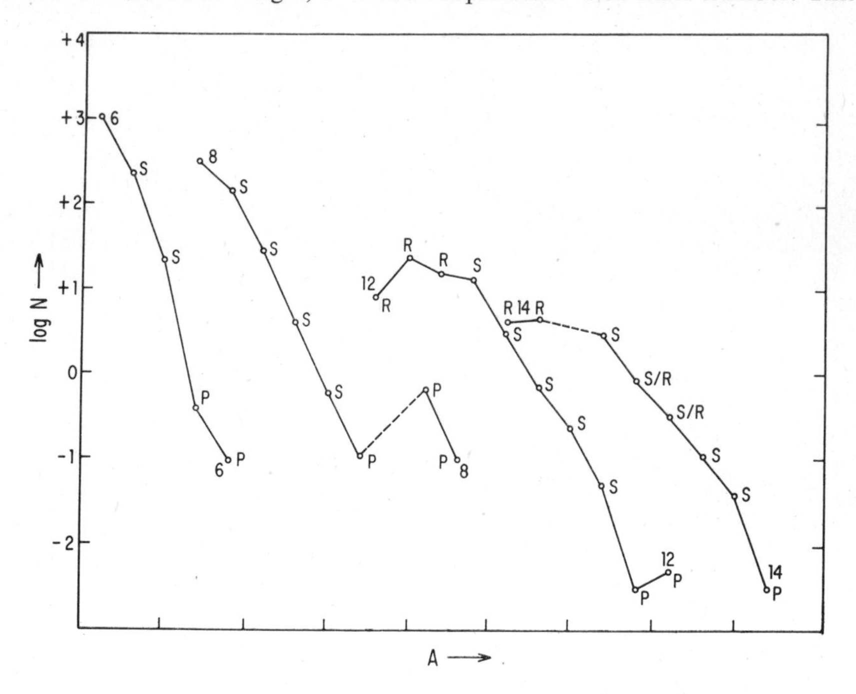

Fig. 3. Abundance values for nuclides with neutron excess 6, 8, 12, and 14 plotted as a function of mass number. The abundance lines are separated horizontally so that their coherent trends can be recognized more easily. It can be seen that relationships exist between the abundances of nuclear species which supposedly originated in completely different processes.

is particularly true for the nuclides with A between 70 and 90. The abundances of nuclear species with a neutron excess of six neutrons, for example, beginning with Ni^{62}, show an approximately exponential decrease with increasing mass number. The same is true for the nuclides with a neutron excess of eight neutrons, beginning with Ni^{64}.

The abundances of nuclei with the same neutron excess are plotted in Figure 3. The abundance lines are separated horizontally in order that the trends be more clearly displayed. The general decrease with mass number is seen to exist within limited ranges for nuclei with $I = 12$ and 14 as well. This pattern is largely independent of the choice of element abundance data, particularly if the requirement is considered that the abundance values for odd mass-numbered nuclear species also show a smooth dependence on mass number.

In Figure 3, the type of process (r, s, or p) is indicated which, according to B^2HF, must have led to the formation of the respective nuclear species. If a given species can have been formed by two processes, both symbols are shown. The striking feature of these curves is that nuclei with the same neutron excess, which have been formed by different, independent processes, have abundance values which lie on the same smooth line. Element synthesis, according to B^2FH, would make it necessary to assume that elements produced from different sources were mixed in just such proportions that conspicuously smooth trends for the abundance distribution resulted or, in other words, that these trends had no physical meaning but were caused by a strange and highly improbable coincidence.

THE r-PROCESS

There is one further result of the recent studies of element abundances which has important implications regarding the r-process. This process is thought to be responsible for the broad abundance peak observed in the rare earth region, $A \sim 128$, correlated with the closed neutron shell at $N = 82$. It is important to note that the usual odd-even effect is not observed in this region. The previous abundance studies by Urey and the author (1956) also displayed an irregularity in the mass region $A \sim 102$.

Preliminary results of the current abundance analyses are displayed in Figure 4. The abundance peak at mass number $A \sim 102$ determined in this analysis is seen to be comparable to the peak at $A = 128$. If one assumes that a component exists around $A = 102$ exactly as large as the one around

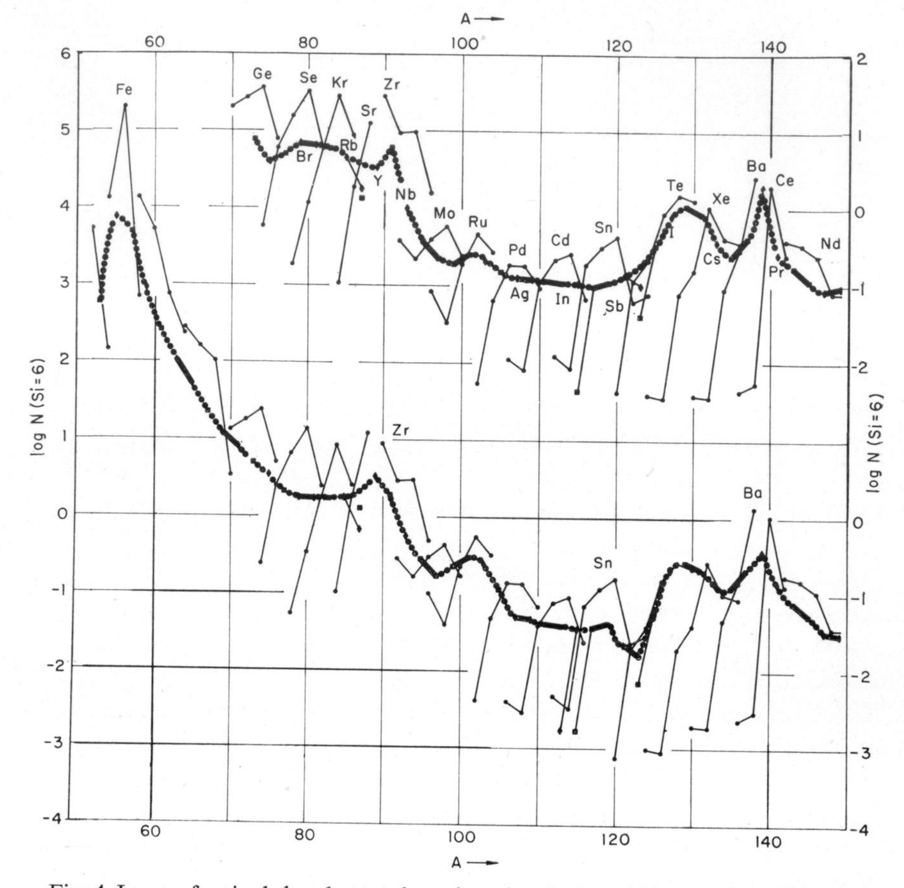

Fig. 4. Log_{10} of revised abundance values plotted against mass number. For comparison the Suess-Urey (1956) data are shown in the upper part of the figure. Points for even A species belonging to the same elements are connected by lines. A heavy dotted line connects the points for the odd mass numbered.

$A = 128$, and if one then subtracts the amounts of this component from the empirical nuclear abundance data around $A = 102$, then one finds for the remaining component the same smooth abundance lines as shown in Figure 3.

The existence of these two similar humps, one that can be associated with the shell closure at $N = 82$ and another that does not seem to be correlated with any nuclear property, eliminates any justification for taking the abundance hump around $A = 128$ as a proof for the existence of neutron capture build-up on a rapid time scale. (Cameron (1963) has argued that the two peaks might result from neutron capture build-up on a rapid time scale, assuming two values for the integrated neutron flux.) Furthermore, the components causing the two humps appear to be mirrored at $A = 115$, and therefore they are suggestive of representing fission products. This is consistent with the absence of an odd-even effect in these components. Also, the nuclides of the components that constitute the two humps are present exclusively in unshielded nuclear species, as in the case of fission products. The curves that can be recognized from nuclear abundance data, however, are narrower than any known fission yield curves; they might, perhaps, represent the irregularities which are caused by the 82-neutron shell and which are superimposed on the much broader fission yield curve. Therefore, the question as to whether or not the two empirical humps in the abundance distribution can be interpreted as caused by the admixture of fission products cannot be decided with certainty, but it is clear that the existence of a correlation between the neutron shell at $N = 82$, and the broad maximum around $A = 128$, can no longer be considered to be proof for the occurrence of r-process synthesis.

CONCLUSIONS

A detailed analysis of nuclear abundance data leaves the impression that no direct evidence exists for element synthesis through neutron *build-up*, but that some indications for element synthesis through *break-up* processes of a polyneutron or of ultra-massive nuclear species can be recognized. Evidently, such *break-up* processes will lead to the formation of hydrogen, of some helium, and of some small residual amounts of heavy atomic species. The composition of this small residual amount will depend on a number of parameters and is difficult to estimate. Preliminary considerations indicate that it may well be possible to understand in a qualitative way the main features of the abundance distribution from such a point of view, at least as long as cosmological boundary conditions are disregarded.

REFERENCES

Burbidge, E. M., Burbidge, G. R., Fowler, W. A. and Hoyle, F. (1957). *Revs. Mod. Phys.* **29**, 547.

Cameron, A. G. W. (1963). *Nuclear Astrophysics*, manuscript based on Yale University lectures.

Suess, H. E. and Urey, H. C. (1956). *Revs. Mod. Phys.* **28**, 53.

Suess, H. E. (1964). *Proc. Nat. Acad. Sci.* **52**, 387.

Observational Evidence Concerning Nucleosynthesis

George Wallerstein

Current theories of nucleosynthesis picture the formation of heavy elements as a continuing process in stellar interiors. This general view was substantiated by the discovery by Merrill (1952) of the presence of the unstable element technetium in the atmospheres of red giant stars. Further, quantitative analyses of a number of high velocity stars (Aller and Chamberlain, 1951; Burbidge and Burbidge, 1956), which are considered to be halo population stars of ages $> 10^{10}$ years, showed these stars to be deficient relative to the sun in such elements as Mg, Ca, Ti, Cr, Mn, Fe, Sr, Y, Zr and Ba (compared to hydrogen) by factors of 10 or 20. Improved analyses of the most extreme of these objects, HD 140283, have shown that the metal deficiency is as great as a factor of 100 (Baschek 1959, Aller & Greenstein 1960). In general, it has been found that the abundance of the heavy elements relative to hydrogen, determined from spectral analyses, is a function of both the time and the location of formation of the star in the galaxy.

Burbidge, Burbidge, Fowler, and Hoyle (1957) have reviewed in detail the role of nuclear reactions in the synthesis of heavy elements in stellar interiors. They attribute the production of elements heavier than hydrogen to eight distinct nuclear burning processes. As the division into these various nuclear burning stages provides a natural outline of this subject, I will consider the observational evidence relating to each of these specific processes.

HYDROGEN BURNING

The hydrogen burning stage constitutes the major portion of the active lifetime of a star. Small mass stars burn hydrogen at a temperature of $1.0\text{–}1.5 \times 10^{7}$ °K by the proton-proton chains while more massive stars, with somewhat higher central temperatures, burn hydrogen by the carbon-nitrogen-oxygen cycle.

Parker, Bahcall, and Fowler (1964) have considered in detail the various possible mechanisms for the termination of the proton-proton chain. They have determined the fractions of H^2 and He^3 consumed by these processes as a function of temperature, providing a means of determining the abundance ratios of the products of this nuclear burning stage. While an observational confirmation of these results is difficult, it is clear that the high abundance of He^3 ($He^3/He^4 \sim 5$) observed in 3 Centauri *A* (Sargent et al. 1961; Jugaku et al. 1961; Bell and Rodgers 1964; Hardorp 1965) cannot result from hydrogen burning by the proton-proton chains. We must conclude that this was formed by some other process.

Caughlan and Fowler (1962) have calculated the abundance of carbon, nitrogen, and oxygen in equilibrium for the CNO cycle under a variety of conditions. A particularly interesting result of these calculations is that the ratio $C^{12}/C^{13} \sim 4$ over a very wide range in temperature, 10^7 to 10^8 °K. There is good observational confirmation that the C^{12}/C^{13} ratio reaches values close to four in carbon stars (McKellar 1948; Climenhaga 1960). This suggests that material which has been processed by the CNO cycle has been carried to the surface of the star by convection, or revealed by mass loss of the outer material.

The results of Caughlan and Fowler (1962) predict a high abundance of nitrogen in stars which have come into CNO equilibrium. Unfortunately the ratio of $C^{12}/C^{13} = 4$ does not prove CNO equilibrium since it can also be reached under non-equilibrium conditions (Caughlan 1965) when the ratio of protons to carbon is small so that most of the protons are used up converting C^{12} to C^{13} and very few protons are left to be captured by the C^{13}. Thus the ratio of carbon to nitrogen in C^{13}-rich stars would reveal whether equilibrium conditions prevailed or not. The C/N ratio would be very difficult to obtain with high accuracy, but order of magnitude work could be done on the C_2 and CN bands of carbon stars.

HELIUM BURNING

Helium burning proceeds via the triple-alpha reaction, resulting in the production of C^{12}. Some buildup of oxygen can take place by alpha capture on carbon, but the uncertainties in the value of this rate (Reeves 1964) make it difficult to determine with accuracy the final ratio of C/O of helium burning. The rate for the triple-alpha reactions has been improved by the experiment of Seeger and Kavanagh (1963).

As helium burning proceeds, the rate of the triple-alpha reaction will fall off much more rapidly with the decrease in the alpha-particle number density that will rate for C^{12} (α, γ) O^{16}. This suggests that if helium burning goes to completion, the major product should be O^{16}. Carbon stars must therefore correspond to stars in which helium burning has not taken place for sufficient time to produce large amounts of oxygen. It is important to note, in this regard, that carbon to oxygen ratios are known only for a very few stars, but that the ratio of C/O must be more than unity in the carbon stars.

The best observational evidence regarding the triple-alpha process is that found in the R Corona Borealis stars. These stars are characterized by a deficiency of hydrogen and a large overabundance of carbon. Two of these stars have now been analyzed-R Corona Borealis (Searle 1961) and RY Sagittarii (Danziger 1965).

Thc spectrum of RY Sagittarii in the vicinity of H_α is compared to that of a normal super-giant Beta Aquarii, in Figure 1. The relative weakness

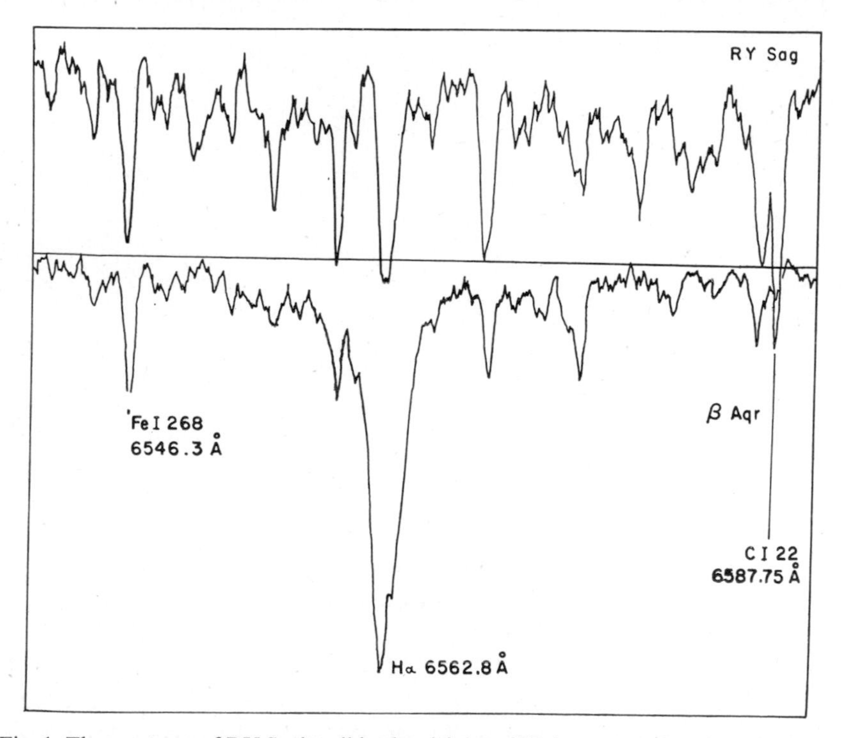

Fig. 1. The spectrum of RY Sagittarii in the vicinity of H_α is compared to that of a normal super-giant, Beta Aquarii.

of the H_α line and the relative strength of the neutral carbon line at λ–6587 Å in RY Sagittarii are quite apparent in this figure. Most of the other spectral features are of roughly comparable strength.

The abundances in RY Sgr relative to β Agr and of RCrB relative to δ CMa are given in Table 1 for a variety of elements. We observe that while carbon is overabundant by approximately a factor of 30, there is no en-

Table I

Log_{10} of Abundances in RCrB and RY Sgr Relative to Ma and βAqr Respectively

Element	RCrB	RY Sgr	Element	RCrB	RY Sgr
C	+1.4	+1.5	V	+0.2	+0.4
N	—	0.0	Cr	+0.1	+0.0
O	—	−0.6	Mm	0.0	0.0
Na	—	+0.7	Fe	−0.1	0.0
Mg	+0.2	−0.2	Ni	—	−0.1
Al	—	+0.2	Zn	—	−0.2
Si	—	−0.2	Y	0.0	−0.5
S	—	+0.1	Zr	0.0	+0.4
K	—	0.0	Ba	0.0	−0.5
Ca	−0.4	−0.5	La	−0.3	−0.3
Sc	+0.4	+0.2	Ce	−0.1	+0.2
Ti	+0.2	+0.2	Sm	+0.2	+0.3

hancement of oxygen. The moderate excess of sodium indicates that there may have been some carbon burning by C^{12} (C^{12}, p) Na^{23}, though we would also expect some enhancement of magnesium in this case. However, as sodium is normally less abundant than magnesium by one order of magnitude, a small admixture of carbon burning products might well result in an enhancement of sodium while affecting the magnesium abundance very little. We also note that there is no enhancement of the heavy elements. a result of considerable interest with regard to the s-process.

THE α-PROCESS

In their original paper BBFH (1957) assumed that the products of helium burning would include large amounts of Ne^{20} and perhaps Mg^{24}. The subsequent nuclear burning process, on this picture, consisted of the photodisintegration of Ne^{20} releasing alpha-particles at $T_9 = 1$, and the successive addition of these particles to the remaining Ne^{20}. In this manner the alpha-particle nuclei Mg^{24}, Si^{28}, S^{32}, A^{36}, Ca^{40} and possibly Ti^{44} and Cr^{48} might be synthesized.

The amounts of Ne^{20} produced during helium burning are now known to be small, due to the small cross section for O^{16} (α, γ) Ne^{20} (Reeves 1964). The dominant processes, following helium burning, will therefore be the destruction of C^{12} by the reaction $C^{12} + C^{12}$ and oxygen burning by the reaction of $O^{16} + O^{16}$ (Cameron 1959, Reeves and Salpeter 1959). While the details of these reactions are not well determined, generally the reaction products should be nuclei with masses in the range $20 \leqq A \leqq 32$, particularly the alpha-particle nuclei Mg^{24}, Si^{28}, and S^{32}. Some production of alpha nuclei past S^{32} may be possible on this picture.

Observational evidence indicating that this process has taken place is relatively poor. There have been no cases reported in which alpha particle elements are in excess by, say, one order of magnitude over the intermediate elements. There are, however, a number of G dwarfs which show small excesses of the α-elements Mg, Si, Ca, and Ti relative to sodium, scandium, and the iron peak elements (Wallerstein 1962). The atmospheres of G-dwarfs could only indicate anomalies formed in the dim and distant past, since we believe they reflect the composition out of which they were formed rather than showing effects of nuclear transformations in thier own interiors.

THE e-PROCESS

The formation of the iron peak elements (V, Cr, Mn, Fe, Co, Ni) is attributed to the equilibrium process. It is found that the solar abundances of these elements can be fitted reasonably well, employing the equations of nuclear statistical equilibrium, for a temperature of approximately 4×10^9 °K and a density $\varrho \sim 10^6$ gm/cm^3. Clifford and Taylor (1965) have performed extensive calculations of these abundance distributions, covering a range in temperature and density. The equilibrium process has recently been studied in detail by Fowler and Hoyle (1965) and by Truran et al. (1966).

The abundances of the various e-process elements can all be determined for the sun (Goldberg, et al. 1960; Müller and Mutschlecner, 1964). These values are determined generally from analyses of the absorption lines of the photosphere. There is an important discrepancy with regard to the iron abundance, as Pottasch (1963) has determined a considerably higher value from a study of the ultraviolet spectrum of the lower solar corona confirming the results of Woolley and Allen (1948). The discrepancy seems to be real and can best be explained by diffusion in the corona and selective escape favoring light elements via the solar wind (Brandt, 1966).

There is evidence that the relative abundances of the e-process elements vary in different stars (Wallerstein et al., 1963). In particular, we have found manganese and vanadium to be deficient with respect to iron in metal-poor red giant stars. The abundance determinations for manganese are somewhat uncertain due to the dependence of this analysis both on the shape of the curve of growth and on the velocity parameter for the sun. It has been found that some of the manganese deficiencies are smaller than had previously been estimated, but are indeed present (Bely-Dubau 1964, Wolff and Wallerstein 1966. The e-process elements in the extremely Metal-poor Star HD 122563 have been rediscussed by Wolff and Wallerstein (1967).

THE s-PROCESS

The s-process consists of neutron capture proceeding on a time scale which is long compared to the typical beta decay half lives of the product nuclei. The s-process path will therefore follow closely the valley of beta stability. The abundance distribution of the products of this nuclear burning stage should peak in the vicinity of the neutron closed shells, where the neutron capture cross sections are systematically small.

In contrast to the other nuclear processes, for which there is little observational confirmation, there is a great deal of evidence available for the existence of s-process synthesis. The presence of technetium (Merrill 1952), an element with no stable isotopes, in the atmospheres of red giant stars constitutes strong evidence that the synthesis of heavy elements by neutron capture is currently taking place (Cameron 1955). The general correlations of cross section and abundance (Gibbons, these proceedings) also tend to confirm the s-process hypothesis.

In addition to the s-type and carbon stars with technetium there are three types of stars which provide good observational evidence regarding the s-process – Barium II stars, CH stars, and some metal deficient stars. The Ba II stars have abundances of metals comparable to the sun while the heavy elements, particularly barium and the rare earths, are in excess. (Burbidge and Burbidge 1957, Warner 1965, Danziger 1965). The CH stars are barium stars which display strong features of CH and appear to be deficient in metals. An analysis of two CH stars (Wallerstein and Greenstein

1964), HD26 and HD 201626, showed them to be metal-poor by factors of 5 and 30, respectively with both showing excesses of Ba, La, Ce, and Nd with respect to iron by factors of approximately 20. Finally, we find in some metal-poor stars a great deficiency of s-process elements, suggesting that the necessary 'seed' nuclei were not available for neutron buildup in stars formed very early in the history of the Galaxy (Wallerstein et al, 1963).

Danziger (1965) has recently compared the observed abundances and those predicted by the total neutron exposure available. The predicted σ_cN curves (the product of the neutron capture cross section and the abundance) for 30 keV neutrons are plotted in Figure 2 for a range of neutron exposures

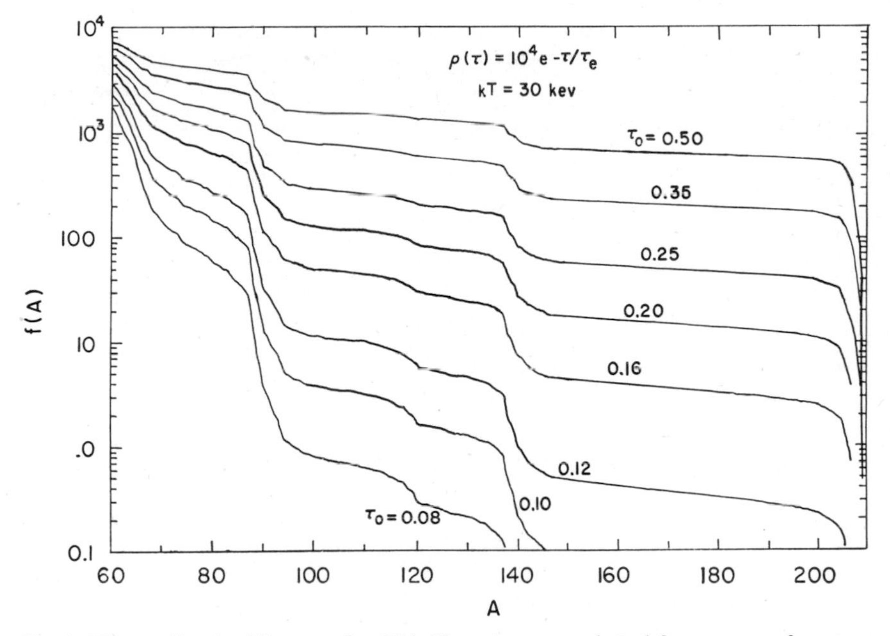

Fig. 2. The predicted σ_cN curves for 30 keV neutrons are plotted for a range of neutron exposures. (Seeger, Fowler, and Clayton, 1965).

(Seeger, Fowler and Clayton 1965). For low neutron exposures there is an abrupt decline in these curves near $A = 90$ (Sr, Y, Zr) corresponding to the closed shell of 50 neutrons, and an even more drastic decline near $A = 138$ (or $n = 82$). As the total neutron exposure is increased these breaks in the σ_cN curve become less pronounced.

The σ_cN curve for the solar abundances is shown in Figure 3. The isotopic abundances are taken from meteoric or terrestrial data. Specific features of

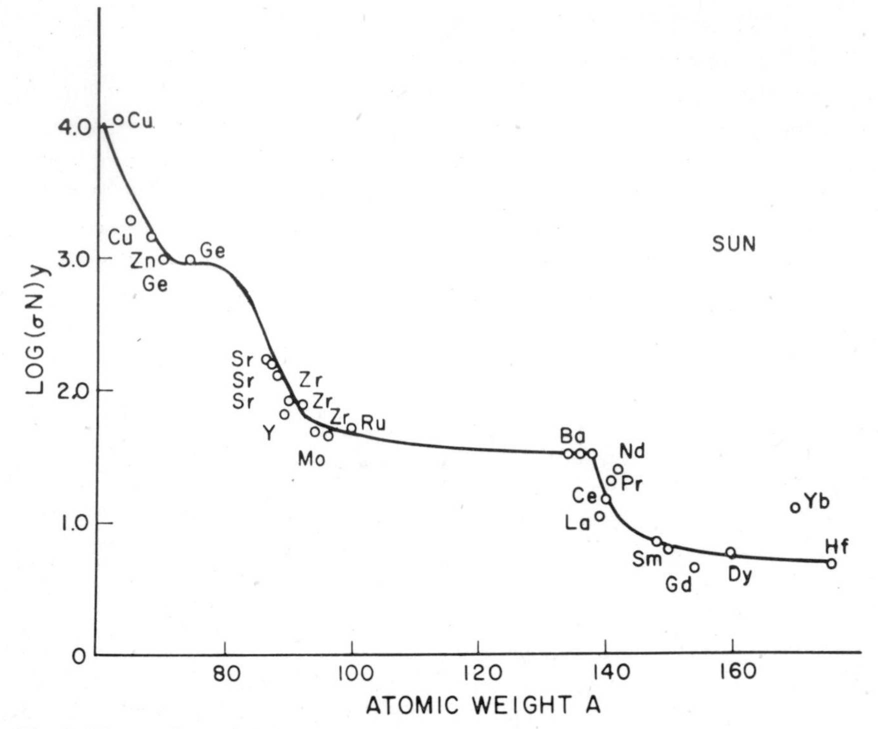

Fig. 3. The product of the neutron capture cross section and the observed abundance is plotted as a function of mass number for the solar abundances.

this curve have been discussed in detail by Dr. Gibbons (these proceedings). We note only that the shape of this curve implies a rather large exposure of seed nuclei.

The σ_cN curve for γ Pavonis, a metal poor subdwarf with a deficiency in metals of about a factor of five relative to the sun, is shown in Figure 4 (Danziger 1965). All of these abundances are measured relative to the sun. The very large break in the curve at $A = 140$ implies a small neutron exposure.

Figure 5 shows the σ_cN correlations determined for the Ba II star HD 116713. Here the metal abundances are roughly comparable with the sun, so the whole curve is elevated with respect to that of the metal-poor stars. These curves do not show a marked decline in the vicinity of strontium, yttrium and zirconium ($A \sim 88$), suggesting that the neutron exposure is larger than that experienced by the solar material. The apparent discrepancies for the elements europium and gadolinium may be real. These

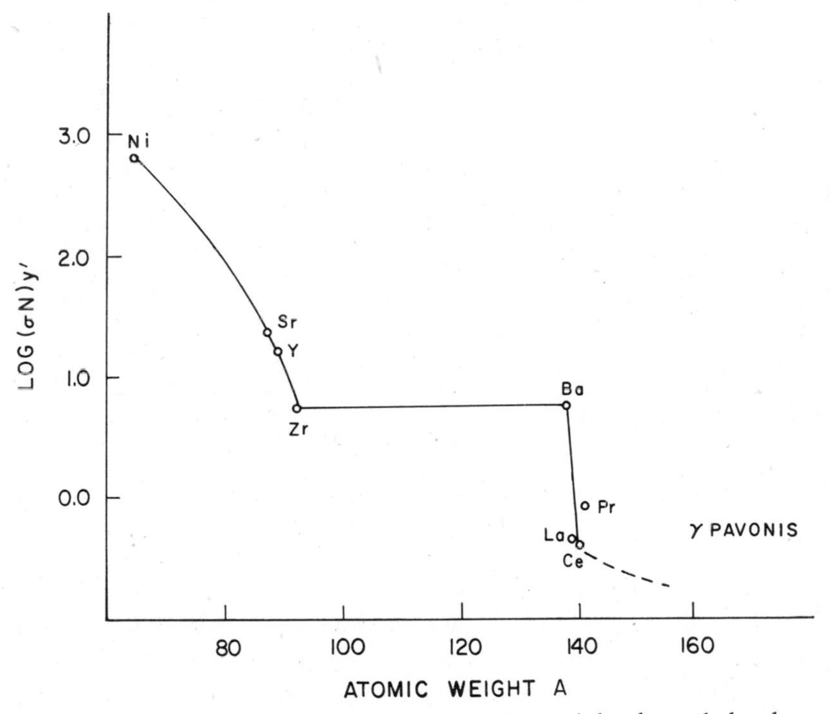

Fig. 4. The product of the neutron capture cross section and the observed abundance is plotted as a function of mass number for γ Pavonis, a metal poor subdwarf.

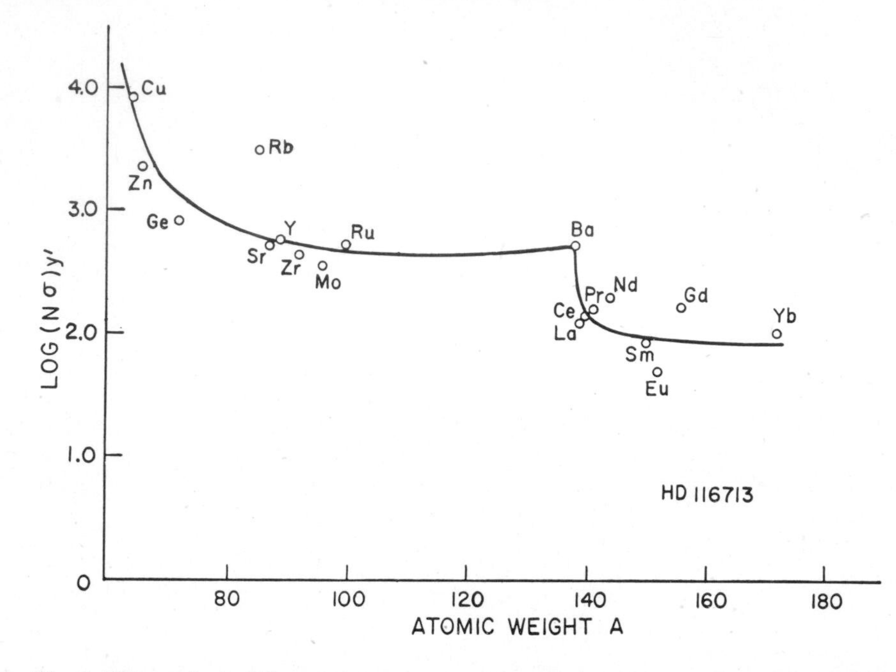

Fig. 5. The product of the neutron capture cross section and the observed abundance is plotted as a function of mass number for the Ba II star HD 116713.

rare earth elements can be formed by r-process neutron capture as well, and we should properly subtract out the predicted r-process contributions. This approach has been taken by Warner (1965) in an analysis of barium stars. In fact, these corrections should be incorporated for all of the elements which can be formed by both processes.

Danziger (1965) has also plotted a mean σ_cN curve for two CH stars, as shown in Figure 6. The metal abundances are somewhat low for these stars, but the total neutron exposure is large as indicated by the large abundances in the rare earth region. Here again we must be careful, as the r-process contributions have not been considered for these elements. The large abundances in this region might imply some rapid neutron capture processing of these elements.

THE r-PROCESS

The r-process of neutron capture, proceeding on a time scale which is short compared to the typical beta decay lifetimes of the product nuclei, results in the production of very neutron rich isotopes which will subse-

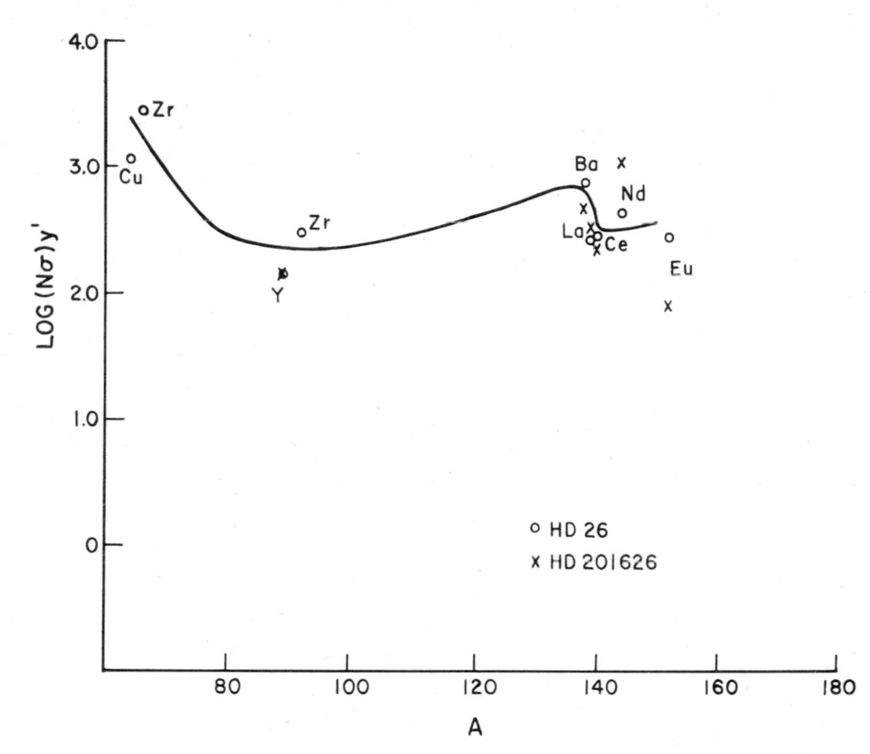

Fig. 6. The product of the neutron capture cross section and the observed abundance is plotted as a function of mass number for two CH stars, HD 26 and HD 201626.

quently undergo beta decays. There is little astronomical evidence available for this process. The only two elements with substantial r-process contributions and that are observable in stars are europium and gadolinium. Solar and meteoric abundance determinations for the heavy rare earths provide a measure of the extent of r-process synthesis as these elements are produced almost entirely by this process. Generally it is necessary to subtract out the s-process contributions in order to obtain a distribution of the products of r-process neutron capture. Recently Fowler, Burbidge, Burbidge, and Hoyle (1965) have discussed the possible importance of the r-process in connection with the magnetic stars.

THE p-PROCESS

The production of some rather rare proton-rich isotopes of heavy nuclei is thought to take place by (γ, n) and (p, γ) reactions proceeding on the products of a neutron capture stage. This is the p-process. The abundances of these rare isotopes are virtually impossible to measure in stars. The evidence for this process must come from studies of solar and meteoric abundances.

THE x-PROCESS

The process (or processes) responsible for the production of the light elements—deuterium, He^3, lithium, beryllium, and boron—is covered by the generic name "x-process". The abundances, of these elements in the sun are low compared to the earth, as they are rapidly destroyed by interactions. with protons. For example, on earth D/H $\sim 10^{-4}$ while in the interior of the sun the equilibrium ratio is D/H $\sim 10^{-17}$. The x-process is usually employed to cover production by spallation but the physical conditions or location of the spallation are difficult to specify.

The deuterium equivalent of the 21 cm. line of hydrogen provides a limit on this ratio in interstellar space, $D/H < 4 \times 10^{-5}$ (Weinreb, 1963). Peimbert and Wallerstein (1965) have set upper limits on the D/H ratio for a number of magnetic stars. The D/H limits in these stars varied from 4.3 × 4.3 × 10^{-5} to 7×10^{-4}. From these results, it was concluded that there is no definite evidence for the presence of deuterium on the surfaces of magnetic stars. We have also searched for deuterium in a number of normal stars, for which we find $D/H < 3 \times 10^{-5}$ (Peimbert and Wallerstein 1965).

The ratio of He^3 to He^4 in the sun is found to be less than approximately 0.02 (Goldberg 1962). In contrast, for 3 Centauri A the ratio is $He^3/He^4 \sim 5$ to 10 (Sargent and Jugaku 1961). As we have mentioned previously, such a high abundance of He^3 cannot result from hydrogen burning by the proton-proton chain.

The determination of the abundance of boron in stars is difficult. Underhill (1958) has reported the possible presence of a weak feature of B III in several 0 stars, but this result is not conclusive. Kohl (1964) has identified two lines of B II in Sirius, although I suspect they may be due to Fe II. I would say there are virtually no observations of boron in stars. A line due to neutral boron has been reported in the sun (Waddell and Slaughter 1966) but it coincides with a CN feature and is probably the latter.

Bonsack (1961) has determined the abundance of beryllium relative to hydrogen in two A stars, α Gem and α Lyr, in good agreement with the value obtained for the sun ($\log_{10}$H/Be = 10). Sirius A (αCMa) was found to be deficient in beryllium by a factor $\sim$100. Bonsack has suggested that some of the material now present on the surface of Sirius A may have come from its binary companion Sirius B. If Sirius B, now a white dwarf, has passed through a red giant phase, the beryllium could have been depelted by a variety of nuclear burning and mixing processes.

The magnetic star $\alpha^2 CVn$ was found to be overabundant in beryllium by a factor of ~100. Sargent et al. (1962) have reported strong Be II lines in some manganese stars, from which they conclude the beryllium abundance is about 100 times that in the sun.

Two very recent studies of Be in F and G stars (mostly of high lithium content) have been completed (Merchant 1966, Conti and Danziger 1966). The last authors found a correlation between the ratio of Li/Be and spectral type, in that the ratio decreases progressively with decreasing surface temperature. They have interpreted this in terms of convective depletion of lithium but not of beryllium. Any convective depletion of lithium in F and G dwarfs puts a severe strain upon theoretical calculations which generally predict convective regions which are not early deep enough except in stars of type K (Weymann and Sears 1965).

The best clues as to the nature of the x-process are provided by the abundance of lithium. Within the last two or three years a great deal of evidence has accumulated on the abundance of lithium in stellar atmospheres but no clear understanding has emerged to explain the origin, preservation, and depletion of lithium in various types of stars. As we shall see, a number of reasonable hypotheses explain certain data but they often fail to explain the observations of other objects.

The abundance of lithium in the sun was first determined by Greenstein and Richardson (1951) and improved by Dubov (1955); Goldberg, Müller, and Aller (1960); and Mutschlecner (1963). Recently Lynds (1965) has used a new scanning spectrometer to trace the solar spectrum at the lithium line at the center of the solar disk. He limits the strength of the lithium line to 1.6 ma, with no evidence for Li^6. Schmahl and Schröter (1965) have derived the lithium abundance from sunspot spectra in which the resonance line is much stronger than in the undisturbed disk and the subordinate line at $\lambda 6104$ is present. They find that the ratio of H/Li lies between 1 and 2.3×10^{11}. They set an upper limit on the ratio of Li^6/Li^7 of 0.10 and find 0.05 to be a likely value.

The T-Tauri stars, which appear to be still in the contracting stage, have very high lithium abundances (Bonsack and Greenstein 1960, Bonsack 1961). It appears that lithium has recently been produced on the surfaces of these stars since they contain a higher concentration of lithium than do the nebulae surrounding the stars or the interstellar medium.

The main-sequence stars in the Hyades cluster provide valuable clues as to the origin of lithium (Wallerstein, Herbig, and Conti 1965). Within the *F* stars the lithium content varies from 100 to less than 15 times the solar lithium abundance (see Fig. 7). It appears that these stars arrived on the

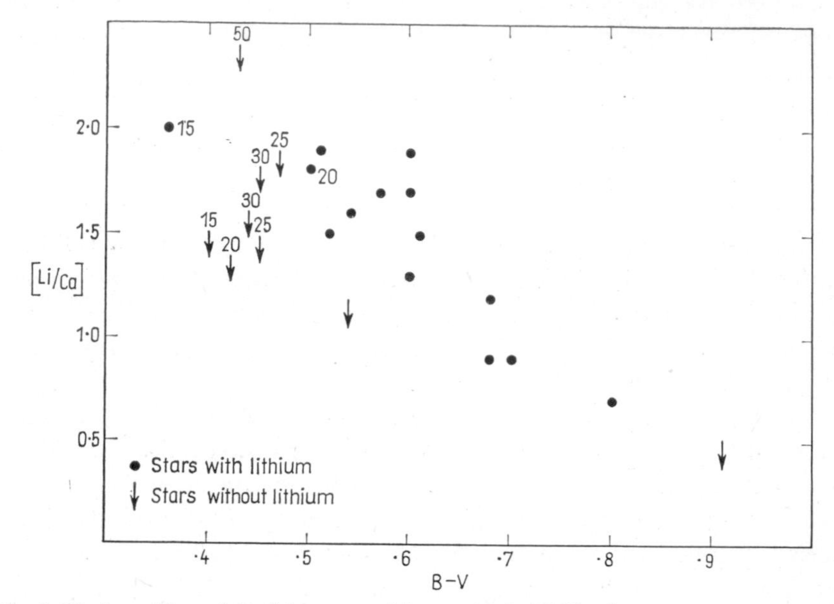

Fig. 7. The logarithm of the lithium to calcium ratios in 23 Hyades stars, compared to the sun, is plotted as a function of the color index.

main sequence with different lithium abundances, due probably to different amounts of lithium production at an earlier stage. Stars later than about type GOV show progressively less lithium. The latter fact has been explained in terms of lithium depletion at the bottom of the hydrogen convection zone during contraction to the main sequence (Bodenheimer 1965).

Herbig (1965) has surveyed 100 F5–G8 dwarfs and found that the percentage with high lithium content steadily decreases with advancing spectral type. He has interpreted this correlation in terms of gradual depletion of lithium at the bottom of the hydrogen convection zone of main-sequence stars, thus implying a correlation between a star's age and its lithium content. The situation is somewhat confusing because Bodenheimer's calculations, which are so successful in explaining the Hyades data, predict a sudden drop in the lithium abundance near type GO while Herbig finds a high lithium content in stars as late as type G8 and low lithium in a few stars as early as

F 5. The latter is understandable if lithium production on stellar surfaces is variable, but the former can only be explained by some mechanism that inhibits convective mixing during the contraction phase or that produces lithium on the surfaces of main-sequence stars.

The predicted correlation between lithium abundance and age has been investigated by Wallerstein (1966) by observing members of visual binary systems which can be dated by methods similar to those used in dating clusters. The resulting correlation is shown in Fig. 8. For the younger stars there is good confirmation except possibly for those stars in the Hyades of

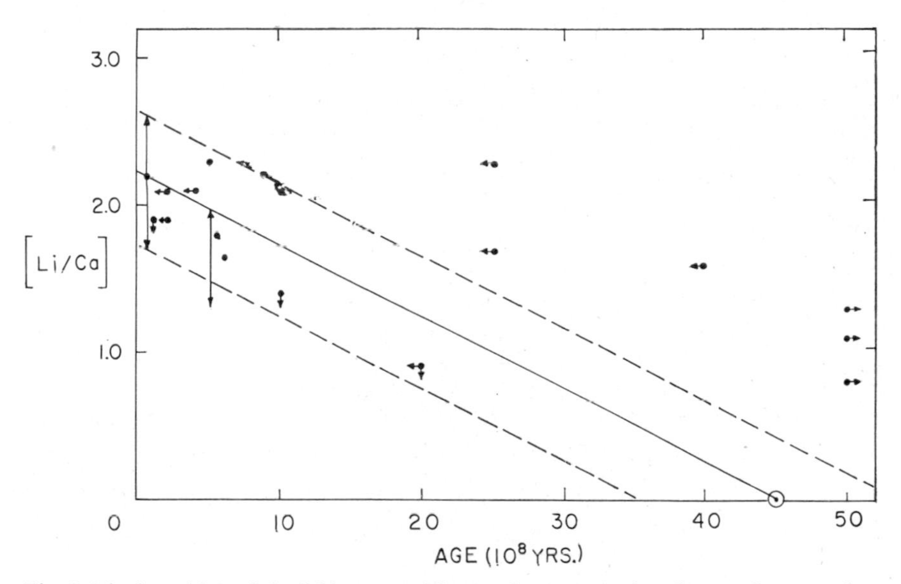

Fig. 8. The logarithm of the lithium to calcium ratio, compared to the sun, for a number of visual binaries, is plotted as a function of the age of the star.

type F which have a low Li content. The serious exceptions are the old stars, including some G-type subgiants that show a high lithium content. Another very old star of type KOIV, δ Eri, has a rather high lithium abundance (Conti 1964). Again it appears that we must modify Herbig's simple age correlation to allow either lithium production on stellar surfaces or lithium preservation by the inhibition of convection.

The possibility that lithium production can take place on stellar surfaces suggests that the magnetic stars with their numerous abundance anomalies and shallow convection zones would be likely candidates for very high lithium content. Most of the magnetic stars are too hot to show the neutral lithium line; nevertheless, all the stars observed for deuterium (Peimbert and Wallerstein 1965a) have been searched for the lithium resonance line. Resulting abundances are shown in Table II. For γ Equ and HR 7575 the lithium abundance is high but no higher than in the T-Tauri stars and some of the F-stars in Figs. 7 and 8. One star in Table I may have a variable lithium line (Wallerstein and Hack 1964). We conclude that the lithium content of Apec stars is not substantially different from that in other young stars.

Table II

Lithium in Peculiar A Stars and Metallic-Line Stars

Star	Type	[Li/Ca]	[Li/H]	Remarks
HR 7575	A5p	2.1	—	Spectrum variable
γ Equ	A7p	1.5	2.2	
β CrB	F0p	<1.3	<1.4	Li possibly variable
HR 710	A4p	<2.2	—	
10 Aql	A4p	<1.7	<1.9	
3 Stars	B9-A0p	—	$\lesssim 3.0$	
9 Stars	ML		$\lesssim 2.0$–3.0	Includes Hyades members

Turning to giant stars we would expect that the lithium abundance should tell us whether or not the outer layers of the star have mixed to depths at which lithium depletion had taken place, i.e., regions of about 3×10^6 °K. For giants of types F, G, and K the agreement of the observations with the predictions of models is quite good. The lithium abundance of K giants is generally considerably lower than that in the sun (Bonsack 1959). Among the *F* and *G* giants Wallerstein (1966) has shown that the "break" in lithium content is found at about type GO so that only *F* giants show high lithium abundances. The system of Capella exemplifies the difference since the F 8III star shows at least 60 times the lithium content of the G 5III component. Iben (1965) has explained the difference in terms of mixing at the tip of the red giant evolution thus implying that Capella G as well as most G and K giants have already mixed and are moving to the left in the color-magnitude diagram.

The M-type and carbon stars do not seem to fit this picture. They are probably more highly evolved than *K* giants yet both groups contain some stars with high lithium content. Merchant (1967) has found the range in the ratio of Li/Ca to extend from twice, to much less than the solar ratio. Similar wide variations in the lithium content of the carbon stars has been reported (Torres-Peimbert, Wallerstein, and Phillips 1964, Torres-Peimbert and Wallerstein 1966). They have noted that the only carbon stars to show a really low lithium abundance are the high velocitiy stars. Observations of about 20 more stars confirms that preliminary correlation. It is very difficult to understand the high abundance of lithium in stars which must have mixed as deeply as the carbon stars seem to have (Cameron 1955).

Concerning the ratio of Li^6/Li^7 we shall be brief. For the earth and meteorites the ratio of Li^6/Li^7 is 1/12.5 with a variability of less than 2% (Krankovsky and Muller 1964). The solar ratio could actually be the same since Schmahl and Schröter (1965) set an upper limit at 1/10 and a likely value at 1/20. For stars, observations by Herbig (1964), Wallerstein (1965), and Wallerstein and Merchant (1965) indicate that the ratio ranges from $\frac{1}{2}$ to near zero but the accuracy is not high since the wavelength difference is only 0.15 Å between the Li^6 and Li^7 features. Without discussing details we merely note that the presence of any Li^6 whatsoever indicates production by nuclear spallation reactions since the proton-proton chain cannot produce Li^6 by any of its closing paths (Parker, Bahcall and Fowler 1964). An extensive review of the nucleosynthesis of lithium, beryllium, and boron has just been published (Bernas, Gradsztajn, Reeves, and Schatzman 1966).

REFERENCES

Aller, L. H. and Chamberlain, J. W. (1951). *Astrophys. J.* **114**, 52.

Aller, L. H. and Greenstein, J. L. (1960). *Astrophys. J. Suppl.*, **5**, 139,

Baschek, B. (1959). *Z. für Astrophys.* **48**, 95.

Bell, R. A. and Rodgers, A. W. (1964). *Observatory*, **84**, 69.

Bely-Dubau, F. (1964). *Thesis*, University of Parts.

Bernas, R., Epherre, M., Gradsztajn, E., Klopisch, R. and Yiou, F. (1965). *Phys. Letters* **15**, 147.

Bernas, R., Gradsztajn, E., Reeves, H. and Schatzman, E. (1966). (Preprint).

Bonsack, W. K. (1959). *Astrophys. J.* **130**, 843.

Bonsack, W. K. and Greenstein, J. L. (1960). *Astrophys. J.* **131**, 83.

Bonsack, W. K. (1961). *Astrophys. J.* **133**, 551.

Brandt, J. C. (1966). *Astrophys. J.* **143**, 265.

Burbidge, E. M. and Burbidge, G. R. (1956). *Astrophys. J.* **124**, 116.

Burbidge, E. M., Burbidge, G. R., Fowler, W. A. and Hoyle, F. (1957). *Rev. Mod. Phys.* **29**, 547.

Cameron, A. G. W. (1955). *Astrophys. J.* **121**, 144.

Cameron, A. G. W. (1959). *Astrophys. J.* **130**, 429.

Caughlan, G. R. (1965). *Astrophys. J.* **141**, 688.

Caughlan, G. R. and Fowler, W. A. (1962). *Astrophys. J.* **136**, 453.

Clifford, F. E. and Taylor, R. (1965). *Memoirs Roy. Astron. Soc.* **69**, 21.

Climenhaga, J. L. (1960). *Publs. Dom. Astrophys. Obs.* **11**, 307.
Conti, P. S. (1964). *Observatory* **84**, 122.
Conti, P. S. and Danziger (1966). *Astronom. J.* **146**, 383
Danziger, I. J. (1965). *Mon. Not. Roy. Astron. Soc.* **130**, 199.
Danziger, I. J. (1965). *Mon. Not. Roy. Astron. Soc.* **131**, 51.
Dubov, E. E. (1955). *Astr. Tsirkular* **159**, 11.
Fowler, W. A. and Hoyle, F. (1964). *Astrophys. J. Supp.* **9**, 201.
Fowler, W. A., Burbidge, E. M., Burbidge, B. R., and Hoyle, F. (1965). *Astrophys. J.* **142**, 423.
Goldberg, L., Müller, E. A., and Aller, L. H. (1960). *Astrophys. J. Suppl.* **5**, 1.
Goldberg, L. (1962). *Astrophys. J.* **136**, 1154.
Greenstein, J. L. and Richardson, R. S. (1951). *Astrophys. J.* **113**, 536.
Hardorp, J. (1965). *Paper presented at the Dec.* **1965** *meeting of the AAS.*
Herbig, G. H. (1964). *Astrophys. J.* **140**, 702.
Herbig, G. H. (1965). *Astrophys. J.* **141**, 588.
Iben. I. (1965). *Astrophys. J.* **142**, 1447.
Jugaku, J., Sargent, W. L. W. and Greenstein, J. L. (1961). *Astrophys. J.* **134**, 783.
Kohl, K. (1964). *Z. für Astrophys.* **60**, 115.
Krankovsky, D. and Muller, O. (O). *Geochim. Cosmochim. Acta* 1625.
McKellar, A. (1948). *Publs. Dom. Astrophys. Obs.* **7**, 395.
Merchant, A. E. (1967). *Astrophys. J.* **147**, 587.
Merchant, A. E. (1966]. *Astrophys. J.* **143**, 336.
Merrill, P. W. (1952). *Science* **115**, 484.
Müller, E. A. and Mutschlecner, J. P. (1964). *Astrophys. J. Suppl.* **9**, 1.
Mutschlecner, J. P. (1963). *Astronom. J.* **68**, 287.
Parker, P. D., Bahcall, J. N. and Fowler, W. A. (1964). *Astrophys. J.* **139**, 602.
Peimbert, M. and Wallerstein, G. (1965). *Astrophys. J.* **141**, 582.
Peimbert, M. and Wallerstein, G. (1965b). *Astrophys. J.* **142**, 1024.
Pottasch, S. R. (1963). *Astrophys. J.* **137**, 945.
Reeves, H. and Salpeter, E. E. (1959). *Phys. Rev.* **116**, 1505.
Reeves, H. (1964). *A Review of Nuclear Energy Generation in Stars and Some Aspects of Nucleosynthesis*, Publications of the Institute for Space Studies, New York.
Sargent, W. L. W. and Jugaku, J. (1961). *Astrophys. J.* **134**, 777.
Sargent, W. L. W., Searle, L. and Jugaku, J. (1962). *Publs. Astron. Soc. Pacific* **74**, 408.
Schmahl, G. and Schröter, E. H. (1965). *Z. für Astrophys.* **62**, 143.
Searle, L. (1961). *Astrophys. J.* **133**, 521.
Seeger, P. A. and Kavanagh, R. W. (1963). *Nucl. Phys.* **46**, 577.
Seeger, P. A., Fowler, W. A., and Clayton, D. D. (1965). *Astrophys. J. Suppl.* **11**, 121.
Sues, H. E. and Urey, H. C. (1956). *Rev. Mod. Phys.* **28**, 53.
Torres-Peimbert, S., Wallerstein, G. and Phillips, J. G. (1964). *Astrophys. J.* **140**, 1313.
Torres-Peimbert, S. and Wallerstein, *G. Astrophys. J.* **146**, 724.
Truran, J. W., Cameron, A. G. W. and Gilbert, A. (1966). To be published.
Underhill, A. (1958). *Publs. Dom. Astrophys. Obs.* **11**, 143.
Waddell, J. and Slaughter, C. R. *Astrophys. J.* (1966) (In press).
Wallerstein, G. (1962). *Astrophys. J. Suppl.* **6**, 407.
Wallerstein, G. (1965). *Astrophys. J.* **141**, 311.

Wallerstein, G., Greenstein, J. L., Parker, R. A. R., Helfer, H. L. and Aller, L. H. (1963). *Astrophys. J.* **137**, 280.
Wallerstein, G. and Greenstein, J. L. (1964). *Astrophys. J.* **139**, 1163.
Wallerstein, G., Herbig, G. H. and Conti, P. S. (1965). *Astrophys. J.* **141**, 610.
Wallerstein, G. and Merchant, A. E. (1965). *Publs. Astron. Soc. Pacific* **77**, 140.
Wallerstein, G. and Hack, M. (1964). *Observatory* **84**, 160.
Wallerstein, G. (1965). *Astrophys. J.* **141**, 311.
Warner, B. (1965). *Mon. Not. Roy. Astron. Soc.* **129**, 263.
Weinreb, S. (1963). *Unpublished thesis, Massachusetts Institute of Technology.*
Weymann, R. and Sears, R. L. (1965). *Astrophys. J.* **142**, 174.
Wolff, S. C. and Wallerstein, G. (1966). *Astrophys. J.* **144**, 419.
Wolff, S. C. and Wallerstein, G. (1967). *Astrophys. J.* (in press).
Woolley, R.v.d.r. and Allen, C. W. (1948) *Mon. Not. Roy. Astron. Soc.* **108**, 292.

Observations of Abundance Anomalies in Magnetic Stars

W. L. W. SARGENT

GROSS PROPERTIES OF PECULIAR A STARS[1]

This paper will summarize the work which has been done during the past few years on abundances of elements in peculiar A and B stars. Provided they have sufficiently sharp lines, these stars are found to have large magnetic fields on their surfaces. Several years ago, Burbidge and Burbidge (1955), and later Fowler, Burbidge, and Burbidge (1955), proposed that these stars had undergone surface nuclear reactions. The point of view, particularly the point of view of the people who proposed the idea originally, appears to be changing, partly as the results of the work done by various people, including Jugaku Searle, and myself. Before discussing the results of abundance analyses of these stars, two uncertainties should be noted. First, it is not known whether the abundance anomalies should be attributed to surface nuclear reactions, or to interior nuclear reactions. Second, it is uncertain whether the process occurs in all A and B stars at some stage in their evolution, of merely to a peculiar fraction of this group.

Some spectra of peculiar A and B stars are shown in Figure 1. The surface temperature of these stars ranges monotonically from 12,000 °K for 41 Tauri down to 8,000 °K for HD 15144. There is a correlation between the color temperature just mentioned and the spectral peculiarity (indicated by the symbols on the right) of the star. The hotter stars are characterized by the most predominant peculiarity, an overabundance of silicon. This is shown in particular by a high excitation line of Si II at 4200 angstroms which occurs in these silicon stars but never in normal stars. This line was first identified as Si II by Bidelman (1962).

[1] For more detail, see Sargent (1964).

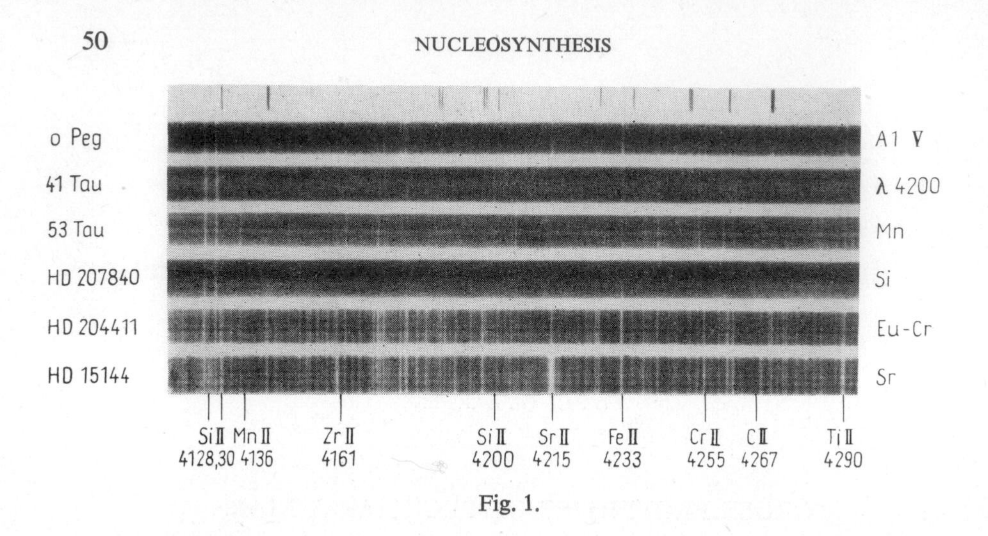

Fig. 1.

The manganese stars are slightly cooler than the silicon stars. Most of the lines in the spectra of the manganese star in Figure 1 are highly excited lines of Mn II, found only in these peculiar stars and never in normal ones.

The cooler stars have strong lines of rare earth elements as their predominant peculiarity. The star HD 204411 has particularly strong lines of chromium, but its temperature is almost the same as the star below it in Figure 1, HD 151144. There are huge differences between the two stars, particularly in the behavior of the lines of strontium and zirconium. The spectrum at the top of Figure 1 is from a normal star O Peg which nearly has the same temperature as these two, but notice the huge enhancement of the number of lines present in the peculiar stars relative to the normal one.

Examples of spectra with deficiencies of a particular element, oxygen, are shown in Figure 2. Spectral classes of the stars are given at the right. There are very weak oxygen lines in α^2 Can Ven, and none in β Cor B. The oxygen lines shown here are the infrared O I triplet $\lambda 7774$, which happen to have conditions of excitation and ionization as the two lines of Mg II which are also shown. Unlike the oxygen lines, these magnesium lines are seen[2] to have more or less the same strength in all the stars shown. This indicates that there is a fantastic effect in some of the peculiar stars, in which oxygen appears to have a negligible abundance. In some of the peculiar stars—in fact, in all of those which have strong lines of manganese—the oxygen lines are normal. The top and bottom spectra in Figure 2 are from normal stars, indicating that the strength of these lines does not vary rapidly with tem-

[2] This is apparent on the original, although not so clear on this reproduction of the spectra.

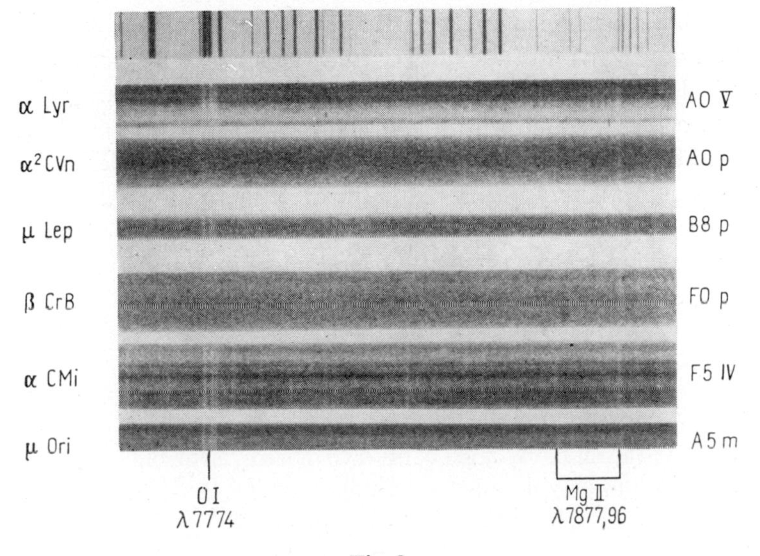
α Lyr
α² CVn
μ Lep
β CrB
α CMi
μ Ori
A0 V
A0 p
B8 p
F0 p
F5 IV
A5 m
O I
λ 7774
Mg II
λ 7877,96

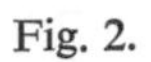
Fig. 2.

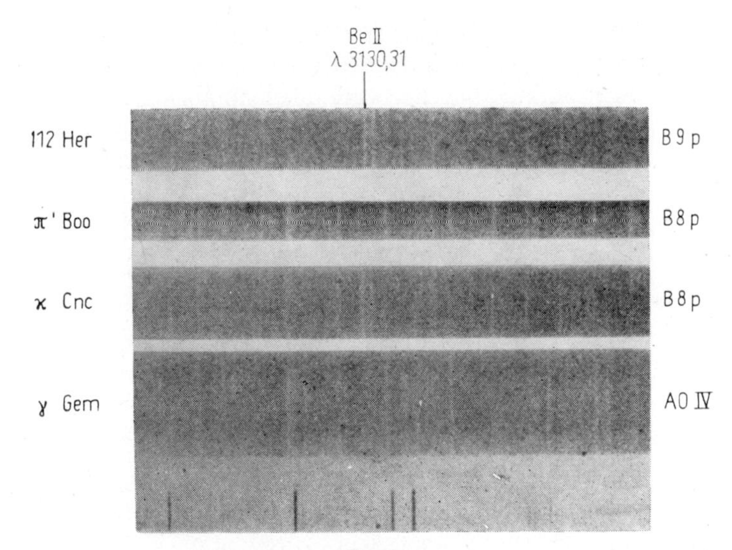
Be II
λ 3130,31
112 Her
π' Boo
κ Cnc
γ Gem
B9 p
B8 p
B8 p
A0 IV

Fig. 3.

perature. Thus, the oxygen deficiency may not be attributed to the stars having a different temperature than the comparison stars.

The erratic behavior of the lines of Be II in some of the peculiar stars is shown in Figure 3. The bottom spectra is from a normal star, γ Gem and possesses a fuzzy feature which is probably Be II blended with lines from other elements. In two of the manganese stars shown, 112 Her and $\varkappa$ CnC, the Be II doublet is very strong, but in the other manganese star, π' Boo, the lines are absent. These three stars have approximately the same temperature. The two beryllium-rich stars have an overabundance of beryllium of roughly 100 relative to the sun.

SUMMARY OF ABUNDANCE RESULTS

Spectra like the ones discussed in this article have been analyzed by various people during the past decade, beginning with the Burbidges' (1955) work on α^2 Can Ven. At present, about a dozen stars have been analyzed in detail. The results of these analyses will be discussed without going into the details of how they were obtained. Although there is general agreement among the authors of these analyses that the observed spectral effects are due to anomalous abundances, it has been suggested[3] that these effects are caused by abnormal atmospheres instead. The viewpoint to be accepted here is that this is not the case, but that the effects are due to abnormal abundances. The evidence in favor of this position is as follows. First, measurement of the temperature of these stars from their spectra indicates that they have the same relationship between color and temperature as normal stars. This may be seen from figure 4 which shows the B-V color plotted against ionization temperature, Θ_{ion}, in the usual spectroscopic notation. The circles refer to normal stars, while the crosses represent certain peculiar stars which have been analyzed in detail.

Second, the curves of growth which one obtains upon analyzing the peculiar stars look quite normal. The turbulent velocities determined appear to be normal also.

Third, the hydrogen line profiles are normal for the color and for the temperature. In fact, there is no evidence from the spectra of these stars that they have any peculiar characteristics *except* abnormal abundances.

[3] Underhill, A. B. (1964) Paper given at the I.A.U. Symposium No. 26 "Abundance determinations in stellar spectra" Utrecht, and in orations delivered at other conferences on miscellaneous topics.

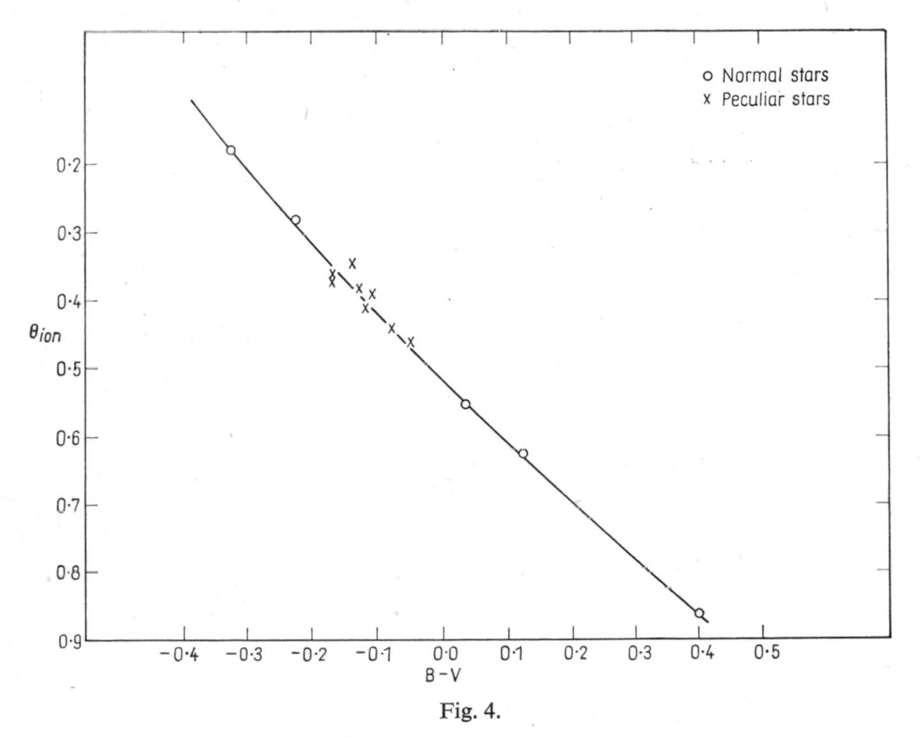

Fig. 4.

A summary of the results of some abundance analyses of Ap stars is shown in Table I. The table lists the logarithms of the ratio of the abundance in the peculiar star to the abundance in the normal star used as reference. Thus +0.7 refers to an overabundance of a factor of 5.

In the first star to be analyzed, α^2 Can Ven, the rare earths were found to be overabundant by a factor of 10^2 to 10^3, and strontium, yttrium and zirconium, which lie between the iron peak and the rare earths, were found to be overabundant by a factor of 10 to 10^2. There is evidence that the whole of the iron peak is overabundant in several of these stars. The matter is uncertain because the opacity in the stellar atmosphere must be known very accurately. Nevertheless, the abnormal strengths of the lines of Fe III, Fe II and Fe I in α^2 Can Ven fairly definitely indicate that iron is overabundant relative to hydrogen. There is an overabundance of silicon in this star by a factor of 10. Calcium is underabundant by a factor of 50; a deficiency in this element is also observed in 53 Tau and HD 133029.

The peculiar behavior of beryllium has been mentioned before; Table I shows that $\varkappa$ Cnc and to a lesser extent α^2 Can Ven have an overabundance of Be while the other peculiar stars do not.

Table I

Summary of abundances (Δ log N) in magnetic stars

Element	α^2 CVn	HD 133029	HD 151199	β CrB	γ Equ	53 Tau	κ Cno	10 Aql
He							−1.2	
Be	+1.2*						+2.0	
C						+0.4		
O	−1.0†			<−1.8†	<−1.7†			−0.9†
Na					0.0			−0.3
Mg	−0.4	+0.1	+0.1	+0.2	+1.2	−0.3‡	−0.8	+0.2†
Al	0.0	+0.3						
Si	+1.0	+1.4	+0.1		+0.5	+0.1‡	0.0	+0.3
P							+2.0	
Ca	−1.7	−1.3	+0.4	+0.1	+0.7	−1.5		+0.2
Sc	−0.2			+0.4	+0.7			−1.6:
Ti	+0.4	+0.4		+0.9	+0.2	+0.6	−0.2	−0.1
V	+0.1	+0.5		+0.4	+0.8			−0.4
Cr	+0.7	+1.0	+0.3	+1.5	+1.3	−0.9	+0.1	+0.4
Mn	+1.2	+1.2	+1.0	(+1.6)	+1.8	+1.8	+1.5	+0.6
Fe	+0.5	+0.6	0.0	+0.8	+0.6	+0.3	−0.3	0.0
Co				+1.0	+1.7			+1.2
Ni	+0.5	+0.4		+0.3	+1.6			−0.5
Zn								−0.4
Ga						+2.4		
Sr	+1.1	+1.2	+1.8	+1.6	+2.6	+1.5		+1.8
Y	+1.3:				+0.6			+0.3
Zr	+1.5	+1.6		+2.0	+0.8			−0.2
Ba	≦0.0		−0.2	+0.7	+1.2			−0.8
Mean Rare earths	+2.8	+2.5	+2.1:	+2.8	+2.2			+1.7:

The overabundance of manganese is almost the only peculiarity or the maganese stars. In Kappa Cancri, the abundance of manganese is up relative to iron—which is normal within the error of measurement—by a factor of 50 to 100. This makes manganese as abundant as iron, whereas the lighter elements, carbon, magnesium, and silicon for instance, are nearly normal. In the manganese stars, very strong lines of the rare earth elements are not observed, but some of the elements beyond the iron peak are overabundant.

In order to summarize this abundance information in a more logical way, the peculiar *A* stars have been divided into three categories. As Dr. Bidelman[4] has pointed out, there are variations within these groups, but the gross characteristics are described by this classification. Table II shows the properties of the three categories: the solicon stars, the manganese stars, and the rare eath stars. As table 2 shows, there is appreciable overlap in the range of B–V or temperature for these groups.

Table II

Properties of the main types of magnetic star

Element	Silicon stars B − V 0.00 to −0.21	Manganese stars B − V −0.02 to −0.11	Rare-earth stars B − V −0.05 to +0.27
He	Very deficient	Normal or slightly deficient	?
Be	Overabundant in some	Very overabundant insome Normal in others	?
C	Very deficient	Normal	?
O	Deficient	Normal	Very deficient
Mg	Deficient	Normal	Near normal
Al	Normal	?	?
Si	Very overabundant	Normal	Near mormal
P	?	Very overabundant in some	?
Ca	Very deficient	Deficient	Near normal
Fe-Peak	Probably overabundant	Normal except in Mn : Fe very high	Probably overabundant
Ga	?	Very overaboundant in some stars	?
Sr, Y, Zr	Very overabundant	Overabundant	Very overabundant
Ba	Normal	? (Not overabundant)	Overabundant
Rare-earths	Very overabundant	? (Not overabundant	Very overabundant

In the manganese stars, helium has a normal or slightly deficient abundance and most of the light elements—carbon, oxygen, magnesium, and silicon—have a normal abundance. Beryllium behaves erratically, being overabundant in some stars. The absolute abundance of the iron peak elements appears to be normal, although the manganese-to-iron ratio is very high. Strontium, yttrium and zirconium are overabundant by factors of 10 to 10^2. Barium is uncertain; although the lines of barium are not visible, it could still be overabundant by small factors.

Bidelman (1960) discovered that in some of the manganese stars phosphorous is overabundant by as much as manganese—a factor of about 10^2.

[4] See Dr. Bidelman's paper in this volume.

It appears that there is a correlation between the behavior of the abundances of phosphorous and beryllium. Only in those manganese stars in which beryllium is found to be overabundant is phosphorus overabundant by a very large factor.

The silicon stars are very deficient in helium; helium lines are not observed in stars of this type which are as hot as a normal B5 or B3 star. It must be concluded that these stars are deficient in helium by a large factor, perhaps 10^2. Again beryllium is erratic, being overabundant in some of the silicon stars but not all. Carbon, like helium, is very deficient in some silicon stars; oxygen and magnesium are slightly deficient. Aluminium is normal, but silicon is overabundant by a factor of 60 to 100. The abundance behavior of phosphorous in silicon stars is unknown. The whole of the iron peak is probably overabundant; strontium, yttrium and zirconium are very overabundant. Barium appears to be normal; this was noted years ago by the Burbidges, but is as yet unexplained. It is difficult to understand theoretically why barium should be normal and the rare earths very overabundant (by factors of several thousand).

The rare earth stars are characterized by extreme deficiencies of oxygen, as Figure 2 shows. Barium is slightly overabundant (a factor of 10) but does not exhibit the large overabundance factors shown by the rare earths (a factor of several thousand). The iron peak is probably overabundant, and strontium, yttrium and zirconium are very overabundant. It should be possible to obtain at least upper limits for the abundances of the more obscure elements as theoretical predictions for these abundances are proposed.

The abundances in 3 Centauri A, a star with a large overabundance of He^3 relative to He^4, are shown in Figure 5. It is slightly hotter than some of the stars just described, but it shares some of their peculiarities. In particular, it has an overabundance of phosphorous similar to that found in maganese stars. However, the manganese-to-iron is normal and beryllium is not overabundant.

The dots with arrows pointing downward in Figure 5 are upper limits for the logarithmic abundance of the element in question. Dots without arrows represent the results of abundance analysis for this star; the solid line is the abundance curve for a normal star. In addition to the overabundance just mentioned, the elements gallium and krypton are very overabundant. This was discovered by Bidelman (1961, 1962b). A glance at the elements beyond the iron peak in Figure 5 illustrates the spectroscopic difficulties encountered in analysis of these stars. For example, the fact that a rubidium line is un-

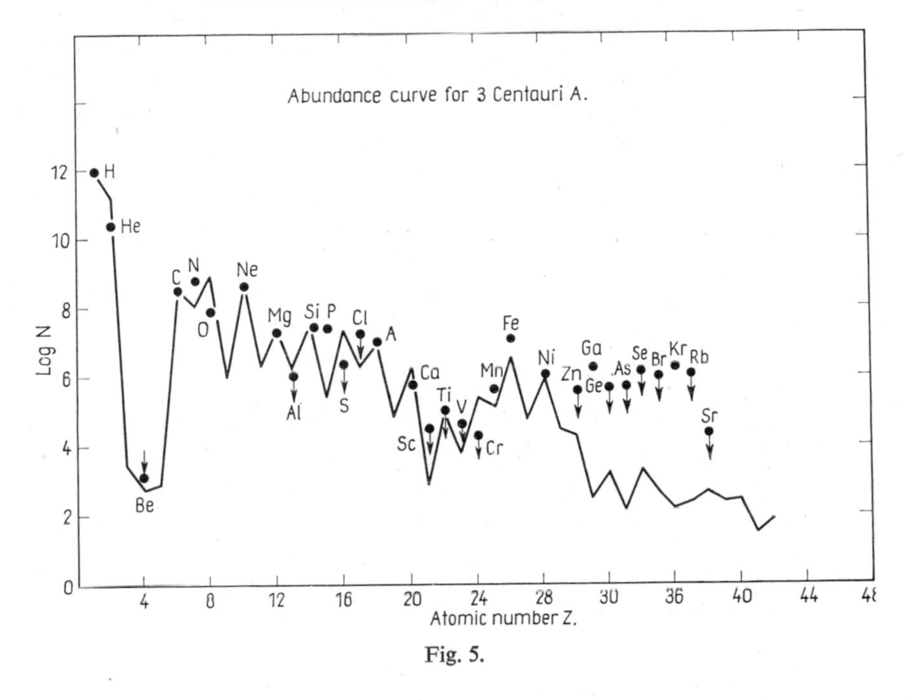

Fig. 5.

observed means that rubidium can still be overabundant by several orders of magnitude. The lines of the rare earths are not present in this star. No analysis has been done to determine how overabundant the rare earth elements might be without their lines being visible. It is quite conceivable that they might be overabundant by a factor of several hundred, with no visible lines, in a star this hot. The star also has deficiencies of sulphur and chromium.

THE LAMBDA BOÖTIS STARS[5]

We now turn to a discussion of the compositions of a group of A-stars which have composition anomalies somewhat different to the magnetic A-stars discussed so far, but which may be related to them.

The Lambda Boötis stars are very rare objects which appear to be metal-deficient A stars. They also appear to be Population I objects, probably evolved young A stars. Fowler, Burbidge, Burbidge and Hoyle (1965) have proposed that the anomalous abundances in the peculiar A and B magnetic

[5] See Sargent (1965).

stars can be explained by a combination of interior and surface nuclear reactions. They also feel that an evolutionary process leads to the peculiar A stars, and that they are not just hot stars with large magnetic fields which behave in a manner different to all the rest of the A-type stars. FB^2H explain the deficiencies of certain light elements by supposing the stellar surfaces have been exposed to a "nucleo-exodus" process by which they are destroyed. Spallation is the most likely mechanism for such destruction, and the high abundance of beryllium, a prominent spallation product, in the atmospheres of these stars is evidence in favor of it. These ideas suggested a search for stars which show evidence of spallation on their surfaces; the author feels that the Lambda Boötis stars may be interpreted in this way.

Lambda Boötis was discovered by Morgan, Keenan and Kelman (1943). They pointed out that this star has a spectral type, as judged from the hydrogen lines, of about AO. This is considerably later than that deduced from the metallic lines, which is another way of saying that the metallic lines are very weak. Several stars have been found to have this property, all of them brighter than fifth magnitude. The other stars are 29 Cyg, Θ Hya, ξ Aur, 2 And, π' Ori, and γ Agr. Parenago (1958) has pointed out that several of these (λ Boo, 29 Cygni, γ Aqr, and 2 And) fall below the main sequence in the color-magnitude diagram for nearby stars. These stars have good parallax determinations. λ Boo and 29 Cyg fall exactly on the zero-age main sequence given by Arp (1958), while γ Aqr and 2 And are, respectively, 1.3 and 1.5 magnitudes below the zero-age main sequence. The remaining stars are between the zero-age main sequence and the evolved main sequence.

There are several reasons for believing that these stars belong to Population I. First, they all have large rotational velocities. The mean rotational velocity for these seven stars is 111 km/sec. It 29 Cyg, π' Ori, 2 And and λ Boo, all of which show a definite metal deficiency, are considered "genuine" λ Boo stars, then the mean rotational velocity for these "genuine" λ Boo stars is 131 km/sec, which is about normal for early A-type dwarfs (140 km/sec). Second, these stars all show a low radial velocity another characteristic of Population I. Third, a large number of these stars are suspected binaries, which is not a Population II characteristic in this region of the H-R diagram. One star, 2 And, has a much fainter visual companion.

As mentioned before, all of these seven stars are brighter than fifth magnitude. Rough statistical analysis indicates that they constitute. one or two percent of all the A-type stars, so that there must be many more stars of this type fainter than fifth magnitude. This fraction must be contrasted with the peculiar A stars which constitute about 12% of all the A-type stars.

The author has suggested that these λ Boo stars are Population I objects which may have undergone spallation reactions on their surfaces as demanded by the theory of FB^2H.

There is a last, speculative piece of evidence about these stars which deserves mention. There is a vissual binary system, ADS-3910, which contains two components of almost the same magnitude. Component A(HR 1753) was classified as B5 V by Slettebak (1963), while component B, 41" distant, was classified as AO pec (?), with the comments: "The lines are somewhat weak for AO III. This could be a λ Boötis star." High dispersion, 8-Å/mm, blue spectrograms of the components of ADS 3910 have kindly been obtained for the writer at the Lick Observatory by Dr. George Wallerstein. Component A appears at this dispersion to be a normal, fairly sharp-lined ($v \sin i \sim 20$ km/sec) B5 V star, while the spectrum of B is practically featureless apart from the Balmer lines. In particular, the lines of He I, which should be easily visible in a star of this color, are not visible. An abundance analysis of this system is in progress. There is little doubt that the stars in this system are physically connected. They have the same radial velocity, although no orbital motion has been detected because they are too far apart. The system ADS 3910 is important in demonstrating that (a) young, Population I stars can exhibit line weakening and (b) this is probably an evolutionary effect since presumably the spectrum of the A component is representative of the interstellar medium from which the stars formed.

If there is a evolutionary connection between λ Boo and peculiar A stars, it must go as follows

normal → λ Boo → peculiar A
stars stars stars

This is because the λ Boo stars have nearly normal rotational velocities ($v \sin i$) while A peculiar stars rotate slowly ($v \sin i \sim 40$ km/sec), with angular momentum being *lost* at some stage.

REFERENCES

Arp, H. C. (1958). *Handb. der Phys.* Berlin/Göttingen/Heidelberg: Springer **51**, 83.
Bidelman, W. P. (1960a) *Publ. Astron. Soc. Pacific* **72**, 24.
(1960b). *Publ. Astron. Soc. Pacific* **72**, 471.
(1961). *Astron. J.* **66**, 450.
(1962a). *Astrophys. J.* **135**, 561.
(1962b). *Astrophys. J.* **135**, 968.

Burbidge, G. M. and Burbidge, E. M. (1955). *Astrophys. J.* Suppl. **1**, 431.
Fowler, W. A., Burbidge, G. R. and Burbidge, E. M. (1955). *Astrophys. J. Suppl.* **2**, 167.
Fowler, W. A., Burbidge, E. M., Burbidge, G. R. and Hoyle, F. (1965). *Astrophys. J.* **142**, 423.
Morgan, W. W., Kenan, P. C. and Kelman, E. (1943). *Atlas of Stellar Spectra.* (Chicago: University of Chicago Press).
Parenago, P. P. (1958). *Soviet Astr. — A.J.* **2**, 151.
Sargent, W. L. W. (1964). *Ann. Rev. Astron. Astrophys.* **2**, 297.
(1965). *Astrophys. J.* **142**, 787.
Slettebak, A. (1963). *Astrophys. J.* **138**, 118.

DISCUSSION

E. M. Burbidge: In your summary[6] of the abundances in the peculiar A stars, you put calcium underabundant by a factor of 50. I believe you have some ideas, do you not, that this may be wrong, and that the ionization temperature is too low?

W. L. W. Sargent: Yes. I think that it is unfortunate that instead of analyzing stars relative to normal stars, people got into the habit of analyzing relative to the first one that was done, so that any error which was committed by Burbidge and Burbidge[7] in the first analysis was amplified in succeeding ones.

B. G. Stromgren: In connection with the ionization temperatures, which elements were used?

W. L. W. Sargent: I think that may graph is wrong at the lower end. At the upper end we used Si III and Si II for all the stars; I think that this is correct.

B. G. Stromgren: At one point you mentioned that the absorption coefficient enters the iron abundance determination. Are there any cases where the overabundance of the metals might lead you to suspect that the metals contribute to the absorption, or is the temperature too high?

W. L. W. Sargent: I think that the temperature is so high that hydrogen always determines the absorption.

E. M. Burbidge: You spoke about evidence for rotation. Do I take it that it is really accepted now that the peculiar A stars are intrinsically slow rotators?

[6] Sargent (1964), Ed. note.
[7] Burbidge and Burbidge (1955).

W. L. W. Sargent: I don't know if everyone accepts this, but people who have studied binary systems recently—Abt. and Deutsch—accept it. I have looked at their evidence and I think that it is probably correct. It was argued at one time that there could be rapidly rotating peculiar A stars whose spectral anomalies would not be seen because the lines would be washed out by the doppler broadening due to the rotation. I think that this idea must be wrong. Some of the abnormalities, particularly in Si and He, would be seen in rapidly rotating stars if they occurred.

Comments on Peculiar A Stars

WILLIAM P. BIDELMAN

I wish to make two points that may further confuse the complicated subject of the peculiar A stars. The first is that I have recently discovered an object, HD 3473, that appears to be greatly overabundant in both silicon and magnesium (Bidelman 1964a, Cowley and Cowley 1965). If this is borne out by subsequent analysis, this star may prove of great interest, as Searle and Sargent (1964) did not find magnesium to be overabundant, in general, in the silicon stars.

The second point that I wish to make is that there are more complexities regarding the rare-earth elements than have yet been discussed in this conference. As you recall, Dr. Sargent pointed out that there are three main groups of peculiar A stars. The coolest of these groups contains the stars whose spectra exhibit very strong lines of the rare-earth elements. Dr. Sargent implied that the overabundances of the rare-earth elements are rather similar among the various stars of the group; however, this is not correct. I believe that one must carefully discriminate between the various types of peculiarities found among the cooler peculiar A stars.

Published data (Sargent 1964) permit an interesting comparison between elemental abundances in the Cr-Eu star β Coronae Borealis and the Sr-Eu star γ Equulei (Morgan, Keenan, and Kellman 1943). One finds that cerium, europium, and gadolinium are substantially more abundant in β Coronae Borealis than in γ Equulei, while just the reverse is true of the elements magnesium, cobalt, and strontium.

Recently I have made a careful, though qualitative, comparison of the spectra of γ Equulei and the star HR 7575 (=HD 188041), which is rather similar to β Coronae Borealis (Bidelman 1964b). Lines of strontium, yttrium, vanadium, cobalt, and neodymium are considerably stronger in the spectrum of γ Equulei, while those of chromium, manganese, cerium, europium, and gadolinium are considerably stronger in that of HR 7575. The differing behavior of the various elements is striking.

This indicates that even in stars which are of rather similar temperature quite different nucleosynthetic processes have occurred. In a very broad and general way one can say that in γ Equulei we have elements present that are largely those produced by the *s*-process, while in HR 7575 (and β Coronae Borealis) the predominent elements are those that are to some extent produced by the r- or p-process. Perhaps there has been more surface spallation in the latter case (Bidelman 1964c).

REFERENCES

Bidelman, W. P. (1964a) "Accurate Spectral Classification By Objective-Prism Techniques," given at the ONR Symposium on "Basic Data Pertaining to the Hertzsprung-Russell Diagram," Flagstaff, Arizona, June 1964. This symposium has been published as Vistas in Astronomy, Vol. 8.

Bidelman, W. P. (1964b). "Line Identifications in Peculiar Stars," given at the I.A.U. Symposium on "Abundance Determinations in Stellar Spectra, "Utrecht, Netherlands, August 1964.

Bidelman, W. P. (1964c) "Conclusions Regarding Nucleosynthesis," given at the I.A.U. Symposium on "Abundance Determinations in Stellar Spectra," Utrecht, Netherlands, August 1964.

Cowley, A. P. and Cowley, C. R. (1965). *Publ. A.S.P.* **77**, 184.

Morgan, W. W., Keenan, P. C. and Kellman, E. (1943). *An Atlas of Stellar Spectra* (Chicago: U. of Chicago Press), p. 19 and Print 32.

Sargent, W. L. W. (1964). *Ann. Rev. Astron. and Astrophys.* 2, 305.

Searle, L. and Sargent, W. L. W. (1964). *Astrophys. J.* **139**, 793.

DISCUSSION

G. R. Burbidge: In all of the analyses that have been done of the rare earths—and there seem to be four stars, from Sargent's table, where a large number of rare earths have been looked at—all of the overabundance ratios are essentially the same.

W. P. Bidelman: I would strongly suggest that this matter needs more looking into. HR 7575 has never been analyzed, and it should be.

Abundances in Metallic-Line Stars

P. S. CONTI

Metallic-line stars were first characterized by the following properties: (1) the K line of Ca II (the strongest line in the solar spectrum) is weak, and (2) the metallic lines are strong, as compared to hydrogen. Subsequently argument developed as to whether the metallic lines were too strong because of anomalous atmospheres or because of anomalous abundances. In this paper the second point of view will be discussed.

The metallic-line stars have several characteristics other than those just mentioned. They are stars with small rotational velocities. Abt (1961) has shown that all, or nearly all, of them are spectroscopic binaries. Recently Abt (1965) has shown that there are no normal stars in the spectral region of the metallic-line stars which are shortperiod spectroscopic binaries. Some of the metallic-line stars are long period binaries and some short period.

Although some of the metallic-line stars have magnetic fields, in general those fields are much smaller than those of the A peculiar stars. The metallic-line stars do have two curious aspects: their atmospheres show a low electron pressure, and their spectra show high turbulence. This has cause some to feel that the metallic-line stars are due entirely to anomalous atmospheres, but I feel that this is not the case.

Figure 1 shows the observed component of rotational velocity, $V \sin i$, for a number of stars, plotted against B-V temperature. These are all stars of the fifth magnitude, and brighter. Normal stars are denoted by open circles and squares. Note that the observed rotational velocity, $V \sin i$, increases toward the left (earlier stars). Note that the slow rotators are predominantly metallic-line (Am) stars and A peculiar stars. There is little overlap between sharp line (that is, slow rotators) normal stars and the metallic-line stars. From this it might be argued that these few normal stars with low $V \sin i$ are being observed along the axis of rotation. I think that this indicates that all stars in this region that are rotating slowly are metallic-line stars.

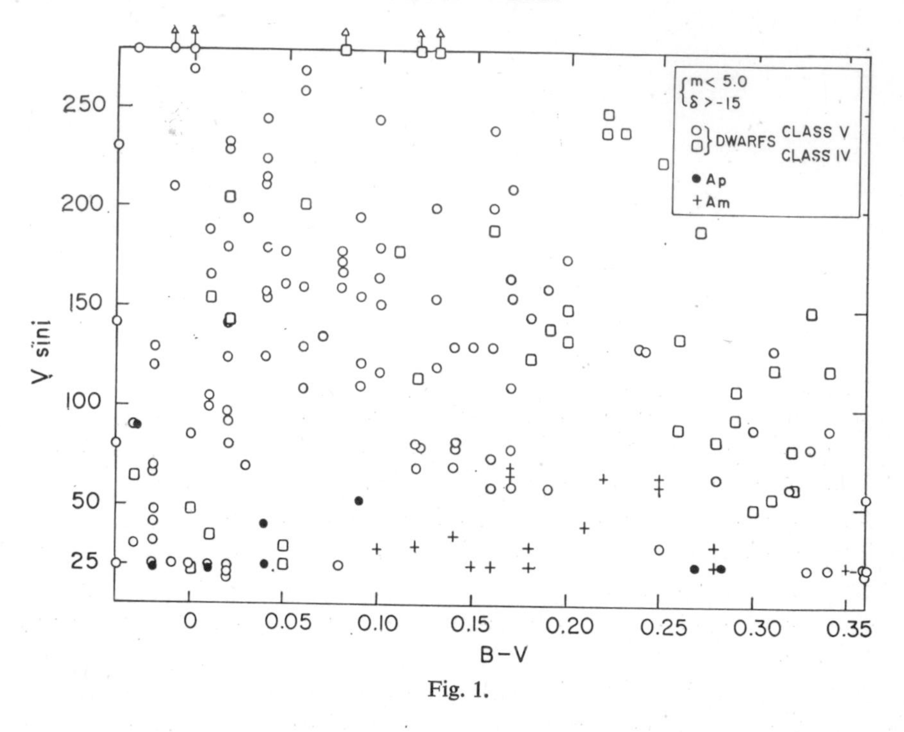

Fig. 1.

There is one metallic-line star which is plotted with B-V of about 0.35 τ UMa). Baschek and Oke (1965) have recently completed photoelectric scans of several metallic-line stars and correlated this with temperature and B–V. These stars are found to have excessive B–V line blanketing from the strong metallic lines relative to normal stars so that they should be moved to the left on this diagram.

Figure 2 shows the Hyades main sequence in the HR diagram. The pluses represent metallic-line stars; the boxes and filled circles are normal stars. The metallic-line stars fall in the diagram with the normal stars, which argues for them being main sequence stars and against them being highly evolved. If they did evolve away from the main sequence, it would be strange if they returned after a time and fitted in the main sequence as nicely as in Figure 2.

In Figure 2 there is a metallic-line star[1] with B–V color less than 0.1, that is, at the extreme left of the diagram. This star is above the turn-off point of the cluster. All cluster main sequence stars with luminosity this high have evolved to red giants, so that this star does not support the argument that

[1] This star is 68 Tauri, and has a magnetic field.

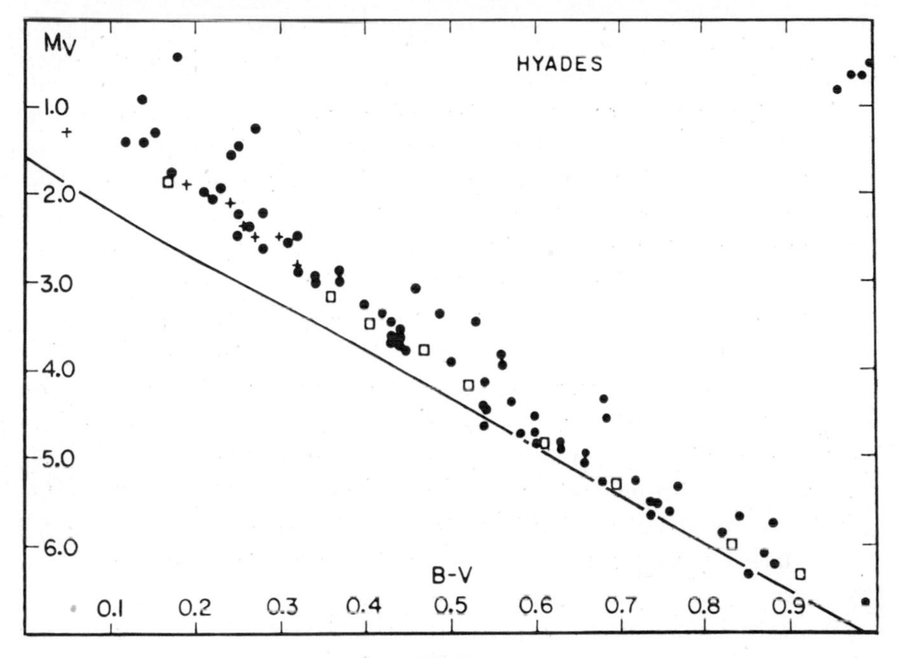

MV
HYADES
1.0
2.0
3.0
4.0
5.0
6.0
B-V
0.1
0.2
0.3
0.4
0.5
0.6
0.7
0.8
0.9

Fig. 2.

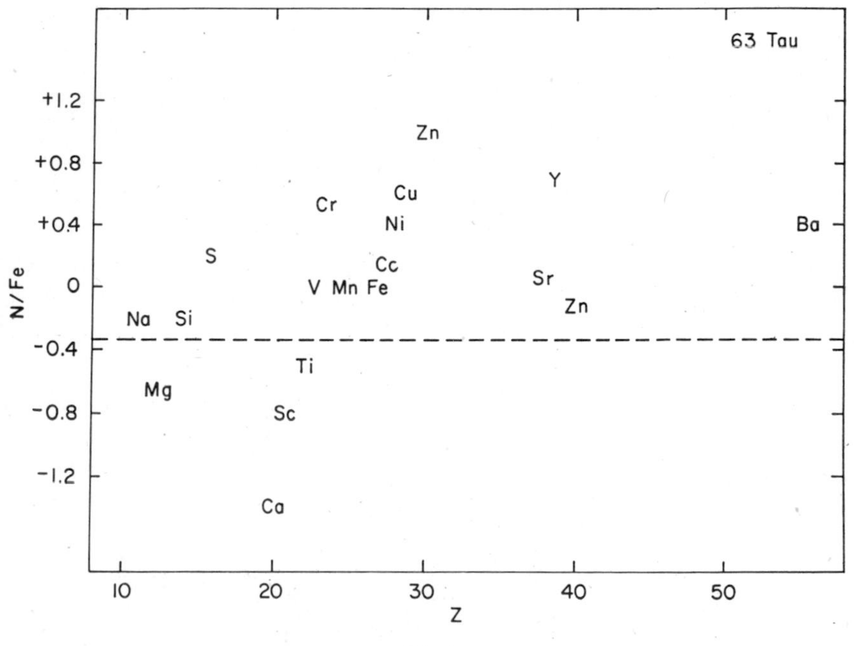

63 Tau
+1.2
+0.8
+0.4
0
-0.4
-0.8
-1.2
N/Fe
Zn
Y
Cu
Cr
Ni
Ba
S
Cc
Sr
V Mn Fe
Zn
Na Si
Ti
Mg
Sc
Ca
10
20
30
40
50
Z

Fig. 3.

metallic-line stars are main sequence stars. Baschek and Oke's measurements of B–V blanketing, mentioned earlier, moves the metallic-line stars to the left in the HR diagram, so that they actually fall *below* the evolved main sequence. Thus the argument for metallic-line stars having evolved off the main sequence and returned is strengthened.

Figure 3 gives the logarithmic abundance anomalies plotted versus atomic number Z for various elements for 63 Tau, a metallic-line star in the Hyades cluster. This star has been studied in detail, using model atmospheres by Van t'Verr Menneret (1963) and myself (Conti 1965). For a normal star at this temperature, the elements would all lie along the dashed line. The zero of the vertical scale is chosen to be at iron. Note the striking underabundance of calcium and scandium, which is characteristic of metallic-line stars. Magnesium is also underabundant: and titanium is a bit low. The other metals are overabundant, that is, their lines are stronger than expected for the surface temperature. The iron is also overabundant with respect to hydrogen.

In Figure 4 abundance anomalies are again shown, this time for another star in the cluster, 81 Tau (Conti 1965). In this case Sc, Ca, Mg and Ti are

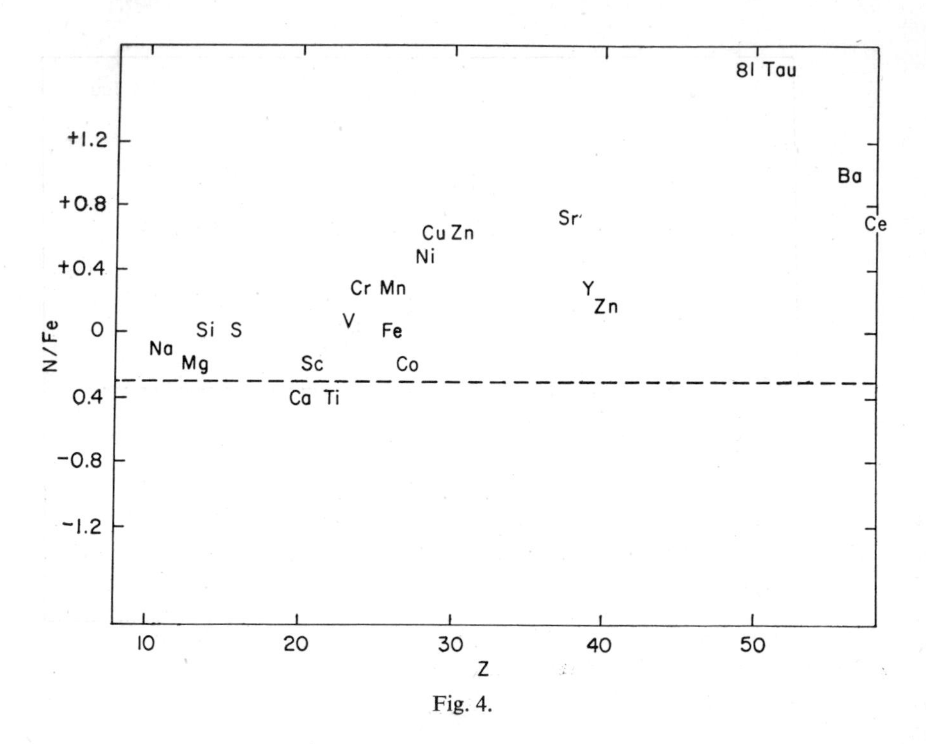

Fig. 4.

roughly as predicted, but all are underabundant relative to iron as before. The rare earths are also overabundant.

Figure 5 shows the abundance anomalies for the star above the main sequence turn-off point for this cluster, 68 Tau (Conti, Wallerstein, and Wing (1965)). Sc is very underabundant; Ca, Ti and Mg are underabundant also. Zinc has only two lines, but they are strong, indicating an overabundance.

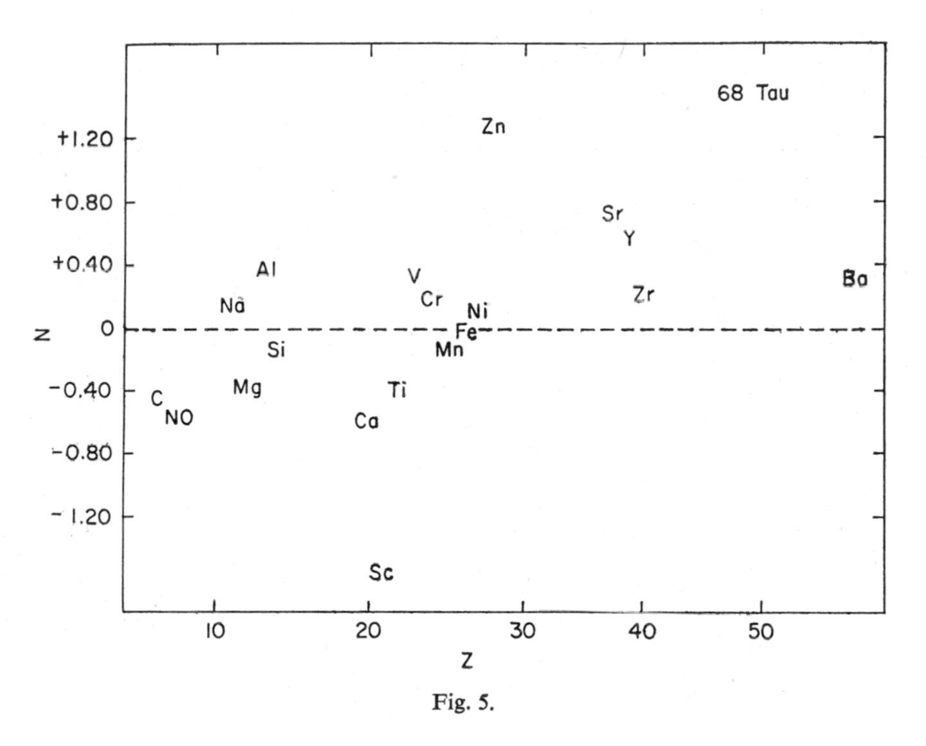

Fig. 5.

In closing consider the confusing case of Sirius. It is not a spectroscopic binary; it has a white dwarf companion.

In a recent detailed analysis (K. Kohl, 1964) found abundance anomalies similar to those shown by 68 Tau in Figure 5. The only difference they found was an overabundance of titanium rather than an underabundance. Scandium was extremely low, and calcium about as shown. In addition, they found carbon, nitrogen and oxygen to be underabundant. These elements are shown to be underabundant in Figure 5, but only a few lines were available.

It is difficult to see what processes other than some sort of nuclear reactions would be able to selectively raise and lower the apparent abundances like this. It is suggestive that Sirius has a white dwarf companion. If the atmospheric material of Sirius A were once inside of Sirius B, then perhaps these anomalies could be explained.

REFERENCES

Abt, H. A. (1961). *Astrophys. J. Suppl.* **6**, 37.
Abt, H. A. (1965). *Publ. Astron. Soc. Pac.* **77**, 121.
Baschek, B. and J. B. Oke (1965). *Astrophys. J.* **141**, 1404.
Conti, P. S. (1965). *Astrophys. J. Suppl.* **9**, 47.
Conti, P. S., Wallerstein, G. and Wing, R. F. (1965). *Astrophys. J.* **142**, 999.
Kohl, K. (1964). Z. *für Astrophys.* **60**, 115.
Van t'Verr-Menneret, C. (1963). *Ann. d. Astrophys.* **26**, 289.

DISCUSSION

Dr. Spiegel: The presence of the high turbulence in these stars is strange because these stars are not expected to be unstable with repect to ordinary convection. The usual explanation for high turbulence in stars that are not very unstable convectively, is that there is some kind of large scale circulation. If there were important circulation—there is no way to explain what might cause it—this circulation would keep the stars mixed and would slow their evolution away from the main sequence. This would explain why they lag behind the normal stars in leaving the main sequence. This circulation would explain how lithium and beryllium are missing at the surface by being destroyed in the hot interior.

On Scandium Production[1]

H. L. HELFER

Summary. It seems likely that Sc^{45} is produced by the *s*-process acting upon α-elements. The excess α-elements observed in a class of G dwarf stars probably have not undergone *s*-processing and this may mean that the α-element enriched material was produced in a comparatively short time compared to the age of the galaxy.

Figure 1 shows the amount of Sc^{45} produced by *s*-processing Ca^{40}, A^{36}, and Ne^{22}. The initial abundances for Ca^{40} and A^{36} were taken from the table of normal abundances of Aller (1961); the initial Ne^{22} was taken equal to the total C, N, and O given in Aller's list. The solid curves give the amount of Sc^{45} produced as function of τ, the neutron exposure (measured in units of 10^{27} neutrons/cm^2); the calculations were performed by the m_k, λ_k method of Clayton et al. (1961). The cross-sections used were generally those of Burbidge, Burbidge, Fowler, and Hoyle (1957) scaled up to give the 30 keV natural isotopic cross-sections of Macklin, Gibbons, and Imada (1963). For Ne^{22}, $\sigma = 1$ mb was assumed.

Since the Clayton et al. investigation showed that $\sim$99.9% of the Fe that is *s*-processed gets neutron exposures with $\tau < 0.6$ (and 2/3 of the Fe gets $\tau \leqq 0.1$) it is clear that in a sample of stellar material that is *s*-processed (if Fe is not abnormally processed and if the elements with atomic weight >22 are present in their normal relative abundances) Sc^{45} is mainly produced from the α-elements present as normal contamination. This conclusion still holds if the CNO cycle has been operative and the N^{14} produced subsequently α-processed to Ne^{22}.

The factor of $\sim$100 overproduction of Sc^{45} for $0.1 < \tau < \sim 0.6$ would imply that only about $\sim$1% of the α-elements are actually *s*-processed; this is comparable to the $\sim$2% of Fe required to be *s*-processed that Clayton et al. have postulated.

[1] This work was supported by a National Science Foundation Grant.

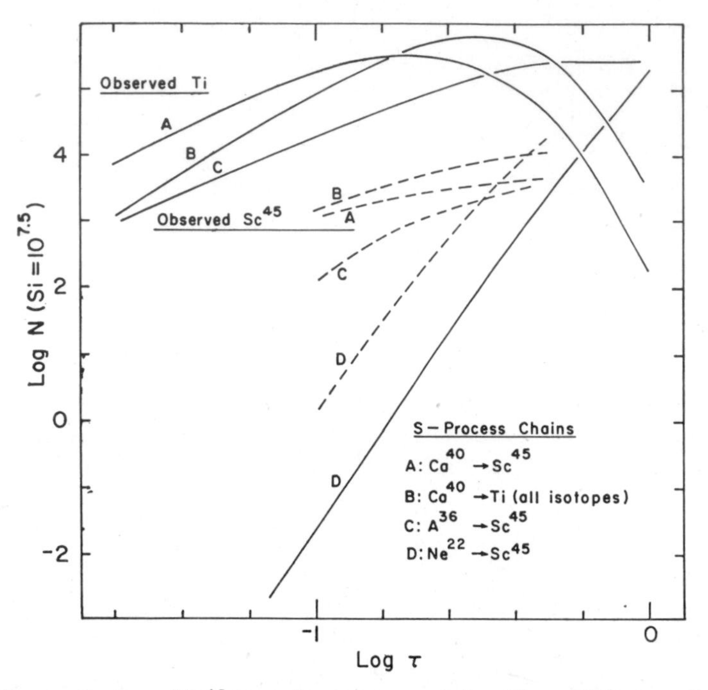

Fig. 1. The production of Sc^{45} by various s-process chains. The solid lines are the results of a calculation based on the method of Clayton et al. (1961); the dashed lines are based upon the exponential formula of Seeger et al. (1965). The two sets of curves have different normalizations (see text).

The dashed curves in Figure 1 represent the production of Sc^{45} as a function of τ_0 according to the recent proposal of Seeger, Fowler, and Clayton (1965) that the number of nuclei exposed to a neutron flux between τ and $\tau + \Delta\tau$, is $\Delta\tau G \exp(-\tau/\tau_0)$. The value of G chosen for the calculation reflects already that only $\sim 2\%$ of the material is to be processed. The authors quoted suggested a value of $\tau_0 \simeq 0.17$ and again within wide limits one finds that Sc^{45} is derivable from the α-elements present as initial contamination.

There is another possibility; Sc^{45} may be produced from a seed nucleus that results from a major combustion process such as Na^{23} resulting from carbon burning. The calculations by Reeves (1963) on neutrino production indicate that higher temperature reaction products resulting from, e.g., oxygen burning can be neglected for there is not sufficient time for the normal *s*-process to be operative on Fe. The *s*-process elements with atomic weights between 22 and 45 possess smaller cross-sections than the *s*-process elements of atomic weight > 56 and consequently should take longer to *s*-process.

For the case of constant cross-section for an s-process chain in which the production of seed nuclei equals the neutron production, one may show that the first fraction x of seed nuclei produced receives an average exposure of $1 - \ln x$ neutrons per seed nucleus. Equivantly, the number of seed nuclei receiving an exposure between τ and $\tau + \Delta\tau$ is $\propto \Delta\tau \exp(-\sigma\tau)$, i.e. the Seeger et al. formula. Since, for carbon burning about 1 neutron per seed nucleus is produced, $\tau_0 = \sigma(Na^{23})^{-1} \sim 0.2$ for the s-process chain originating at Na^{23}. If, after hydrogen and helium burning, $\sim \frac{1}{4}$ of the stellar material is initially C^{12}, then effectively one raises the dashed $Ne^{22} \rightarrow Sc^{45}$ curve in Figure 1 by 2.4 in $\log N$, yielding $\log Sc^{45} \sim 4.5$.

However, one must also ask what happens to the s-processing of any other element, such as Fe^{56}, present as initial contamination. Crudely, out of every Δn neutrons ($= \Delta Na^{23}$) produced, the number of neutrons captured per nucleus of species i is

$$\frac{\sigma(i)\,\Delta Na^{23}}{(\sigma N)_{C^{12}} + (\sigma N)_{O^{16}} + (\sigma N)_{Na^{23}} + \cdots}$$

or, at the end of the production of neutrons and of Na^{23}, the number of neutrons per nucleus of species i is, roughly:

$$\approx \frac{\sigma(i)}{\sigma(Na^{23})} \ln\left[1 + \frac{\text{final}}{\text{initial}} \frac{(\sigma N)_{Na^{23}}}{(\sigma N)_{C^{12}} + (\sigma N)_{O^{16}}}\right] \sim 2{\cdot}5 \frac{\sigma(i)}{\sigma(Na^{23})}\,.$$

This value must be equal to $\sigma(i)\,\tau$, hence $\tau \sim 0.5$. This value of τ is quite high for it would produce too strong an s-processing of Fe. However, we note that the production of Sc^{45} from α-elements present as initial contamination is still about ten times the Sc^{45} production from Na^{23}.

The results depend upon the uncertainty in the cross-sections, of course, but use of the cross sections for 30 keV for the carbon burning stage is justifiable. For these light weight nuclei, the experimental results of Macklin et al. (l.c.) indicate a very strong energy dependance of the cross-sections, favoring the low energy cross-sections used.

Consider now the observed group of G-dwarf stars possessing an overabundance of the α-elements (Wallerstein 1961; Helfer 1964). These stars possess normal values of [Sc/Fe], not abnormal values of [Ba/Fe], excesses of α-elements, and galatic orbits with small perigalactic distances. The normal value of [Sc/Fe] observed in these stars, in light of the above discussion, seems to imply that the excess amount of α-elements added to the material incorporated in these stars has suffered no s-processing. The *excess* amount of α-elements is equal to about two-thirds of the normal amount of α-elements expected in these stars and corresponds to about a hundred times the amount of α-elements contained in three extremely metal deficient red giant stars previously studied. This appears to be much more than we would normally associate with the α-element production of one generation of massive stars. The excess amount of α-elements is at least one fifth the amount of α-elements in the sun and may even be more if these stars actually have an excess of He and are hydrogen deficient. The α-elements are now presumed to be formed in the immediate pre or post supernova stages of a star. Since ordinary carbon burning can s-process this α-element material, it appears as if a very large number of supernovae may have occurred in a relatively short period of time ($<$ a galactic mixing time) in the inner regions of the galaxy*, just before these α-rich stars were formed.

Miss Stephania Zalitacz ably performed the calculations referred to.

REFERENCES

Aller, L. H. (1961). *The Abundance of the Elements* (Interscience Press; New York).

Burbidge, E. M., Burbidge, G. R., Fowler, W. A., and Hoyle, F. (1957). *Rev. Mod. Phys.* **29**, 547.

Clayton, D. D., Fowler, W. A., Hull, T. E. and Zimmermann, B. A. (1961). *Ann. Phys.* **12**, 331.

Helfer, H. L. (1966). Stellar Evolution, R. F. Stein and A. G. W. Cameron, editors: Plenum Press, New York.

Macklin, R. L., Gibbons, J. H. and Imada, T., 1963. *Phys. Rev.* **129**, 2695.

Reeves, H. (1963). *Astrophys. J.* **138**, 79.

Seeger, P. A., Fowler, W. A. and Clayton, D. D. (1965). *Astrophys. J. Suppl.* **11**, 121.

Wallerstein, G. (1961). *Astrophys. J. Suppl.* **6**, 407.

* *Note added in Proof*:—This restriction to the inner regions of the Galaxy is probably not warranted. Since this paper was presented, several metal-deficient stars with nearly circular galactic orbits showing the same relative overabundance of the α-elements have been found.

Thermonuclear Reaction Rates: On the Nucleosynthesis of the S and P Elements

HUBERT REEVES

One of the problems associated with the build-up of heavy elements by neutron capture is: where do you find the neutrons when you need them? I shall discuss here some rough attempts to combine some of the nuclear physics of neutron producing reactions with models of stellar evolution computed by other investigators. It would be desirable to connect the two by means of a model which would put in explicitly the neutron-producing reaction rates as a function of time, and then, follow the resulting neutron flux. This would have little effect on the calculation of the model, but would be invaluable in yielding integrated flux-times with which to follow *s*-process paths. As an exploratory work, however, it is interesting to investigate the effect of some of the neutron reactions.

It was once thought that strong neutron sources, delivering several tens of neutrons by seed nuclei, (usually iron), were needed to build the heavy nuclei. With the work of Clayton and collaborators (1961), it was realized that the distribution of heavy elements has been formed by the action of a large number of weak sources. Thus it will be necessary to examine every neutron source which occurs during stellar evolution, even if its yield in neutrons is actually very small. Hence, I shall consider compound reactions of the form,

$$A + B \rightarrow C^* \rightarrow D + \text{neutron, where } C^* \text{ is}$$

the intermediate excited state formed by the reaction between the species A and B. The rate for the reaction is taken to be of the form

$$p = n_A n_B \langle \sigma v \rangle_{AB} \left(\frac{\bar{\Gamma}_n}{\Gamma} \right) F \text{ per sec-cm}^3,$$

where Γ_n and Γ are, respectively, the neutron and total widths, and F is a factor which is unity for exothermic reactions and which is given for endothermic reactions, by

$$F = \int_{E_t}^{\infty} \sigma(E)\, n(E)\, v\, dE \Big/ \int_{0}^{\infty} \sigma(E)\, n(E)\, v\, dE$$

where E_t is the threshold energy. The number of neutrons emitted per collision is then given by

$$g = \left(\frac{\bar{\Gamma}_n}{\Gamma}\right) F = \frac{N_n}{N_a} = \frac{\text{neutrons emitted}}{\text{concentration of species } A}$$

(See for example, Reeves 1964).

If one knows the isotopic distribution of the gas, then the number of neutrons available to form heavy nuclei can be computed.

Other parameters we will need are the neutron density, (n_n), given by:

$$n_n = n_A n_B \langle \sigma v \rangle_{AB}\, F \left(\frac{\bar{\Gamma}_n}{\Gamma}\right) \Big/ \sum_i n_i \langle \sigma v \rangle^i_{n\gamma}$$

where the sum over i refers to neutron absorbing nuclei, and $t^i(n, \gamma)$, the lifetime of a given element against (n, γ) reactions, which is given by:

$$t^i\,(n, \gamma) = \frac{1}{n_n \langle \sigma v \rangle^i_{n\gamma}}$$

For a given reaction, the irradiation will be characterized by

$$\Delta T = \int n_n\, v\, dt$$

where n_n, is given by the above. An example of a neutron producing reaction is $C^{13}(\alpha, n)\, O^{16}$ which has been suggested by Cameron, and used lately by Wallerstein and Greenstein (1964) as a way of explaining the peculiar abundances found in CH stars, where mixing of the hydrogenic outer layers into the helium core during the helium flash may produce C^{13}. The temperature at which this would occur is close to 10^8 °K, a typical temperature for the helium flash, more or less independent of stellar mass.

Another interesting set of reactions follows the completion of the CNO hydrogen burning stage. Here we have the sequence

$$N^{14}(\alpha, \gamma)\, F^{18}(e^+, \gamma)\, O^{18}(\alpha, n)\, Ne^{21} \quad \rightarrow Q = -700 \text{ KeV}$$

$$O^{18}(\alpha, \gamma)\, Ne^{22}(\alpha, n)\, Mg^{25} \rightarrow Q = -482 \text{ KeV}$$

$$O^{18}(\alpha, \gamma)\, Ne^{22}(\alpha, \gamma)\, Mg^{26}, \text{ exothermic}$$

The rates for these reactions are shown in Figure 2. I shall defer discussion of the reaction leading to Ne^{21} until later and focus attention on the

$Ne^{22}(\alpha, n)$ reaction ($Q = -482$ KeV). The rates for this reaction were based on an optical model calculation. The accuracy of the result was analyzed, assuming a Porter-Thomas distribution of level widths. From the latter, we expect the rate of the (α, n) on Ne^{22} reaction to be good to within a factor of three. Results of the calculations on the number of neutrons emitted relative to CNO and iron (assuming CNO to iron ratio of seventy, from B^2FH) are summarized in Table I where T_8 is the temperature in units of 10^8 °K.

Table I

T_i	1.5	2.0	2.1
N_n/N_{CNO}	0.07	0.4	0.5
N_n/N_{Fe}	4	30	35

We see that significant neutron fluxes are obtained, providing that the temperature exceeds 2×10^8 °K, corresponding, of course, to the region where the Gamow peak tends to coincide with the threshold energy. To find values for the other parameters mentioned before, we must turn to an actual model. The model chosen here is from Hoffmeister et al. (1964) for a 7 $M_\odot$ star whose evolutionary track is shown in Figure 1. For a temperature of 2×10^8 °K, a value for n_n of 10^5 per cm^3 is inferred from the model, and if a typical cross section of 100 millibarns is assumed for Fe, then the lifetime against neutron absorption is of the order of 10^{10} sec. (close to a thousand years), which is clearly in the *s*-process ballpark. The ΔT value is 0.1×10^{27}.

The neutron sources at 1.5×10^8 °K, which is about a factor of one hundred less than the peak values, should not be ignored because, again, these may be important for certain type of stars.

The corresponding numbers for the $O^{18}(\alpha, n)\,Ne^{21}$ reaction are given in Table II, where the temperature must be about 3×10^8 °K in order to obtain a significant number of neutrons released. Such a temperature may be reached during the helium flash for the duration of a few seconds.

We consider now the carbon burning stage with the reaction

$$C^{12} + C^{12} \rightarrow Mg^{23} + n$$

Table II

T_8	2.0	3.0
N_n/N_{CNO}	10^{-3}	0.05
N_n/N_{Fe}	0.07	3
ΔT	2×10^{24}	0.1×10^{27}

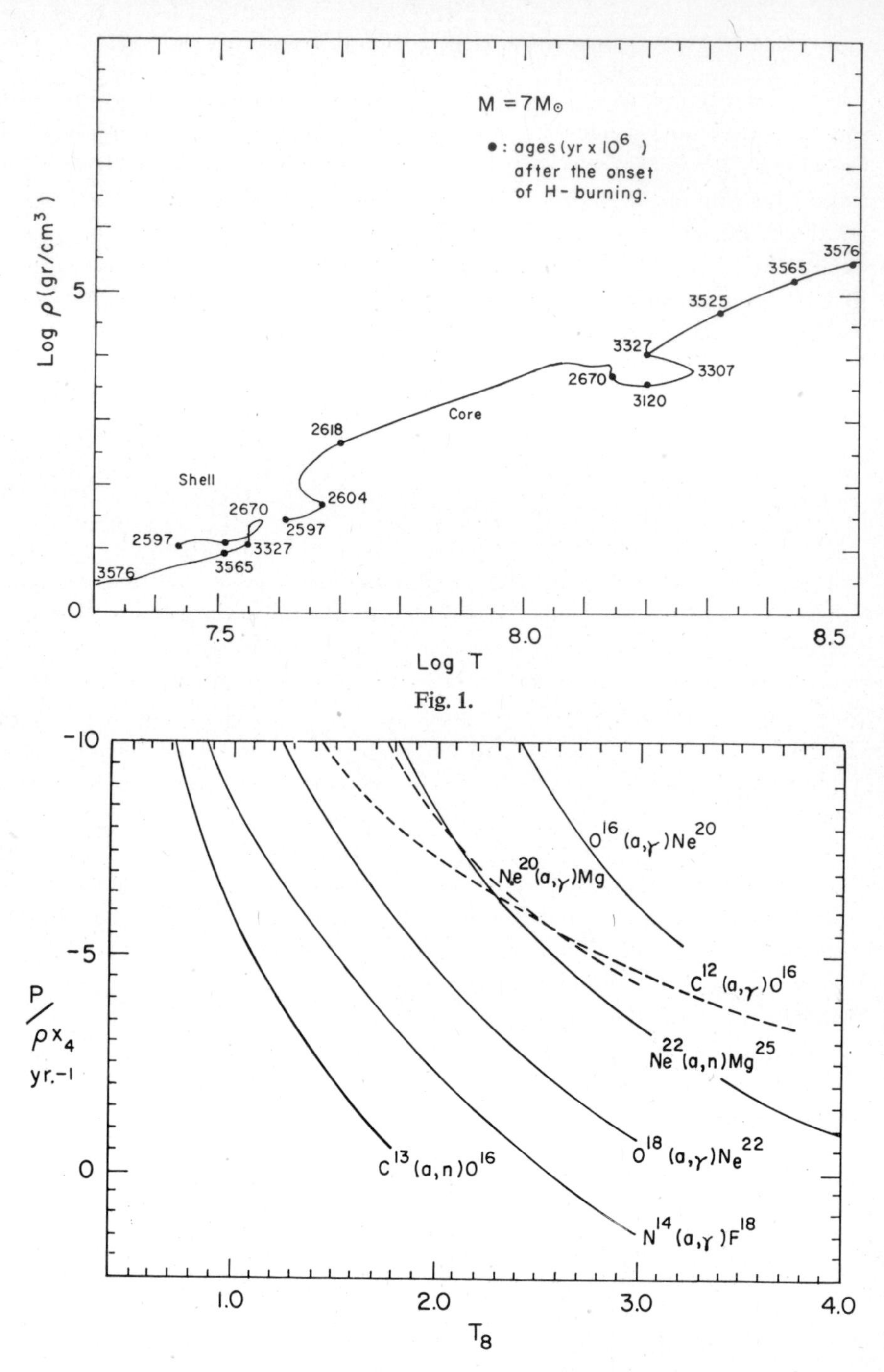

M = 7M⊙
●: ages (yr x 10^6)
after the onset
of H- burning.
Log ρ (gr/cm³)
Log T
Shell
Core
3576
3565
3525
3327
3307
3120
2670
2618
2604
2597
3327
3565
2597
2670
3576
0
5
7.5
8.0
8.5
P/ρX₄ yr.-1
T₈
O16 (α,γ)Ne20
Ne20(α,γ)Mg
C12 (α,γ)O16
Ne22 (α,n)Mg25
O18 (α,γ)Ne22
C13 (α,n)O16
N14 (α,γ)F18
-10
-5
0
1.0
2.0
3.0
4.0

Fig. 1.

Fig. 2.

This reaction was originally thought to be unimportant because the temperatures during the carbon burning stage were taken to be in the range 6–7 × 10^8 °K. Now, with the inclusion of neutrino emission, the temperatures rise to the 8–11 × 10^8 °K level and the reaction becomes important. (See, for example, Figure 3 from a calculation of Hayashi, et al. (1962)).

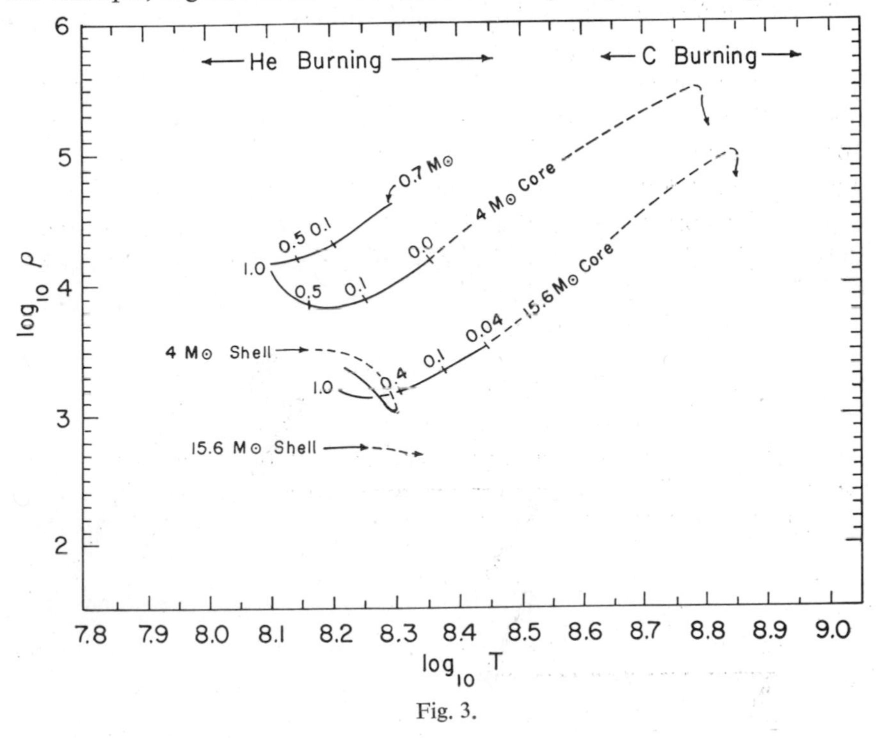

Fig. 3.

From an optical model calculation, good to within about a factor of three, coupled with the evolutionary track of Figure 1, we obtain the values given in Table III.

It is interesting to note that the lifetime against neutron capture $t(n, \gamma)$ decreases with temperature by such large factors. Thus, what may be *s*-

Table III

T_9	0.8	0.9	1.0	1.1
$N_n/N_{C^{12}}$	10^{-3}	3×10^{-3}	6×10^{-3}	10^{-2}
N_n/N_{Fe}	2	6	12	20
$\log (n_n)$	7.6	9.2	10.4	11.6
$\log [\sigma(n, \gamma)\, t(n, \gamma)]$ barn-sec.	7.9	6.3	5.1	3.9

process at 9×10^{8}°K, may not be *s*-process at 1.1×10^{9}°K, where the lifetime is a few hours. This yields the possibility of different branches, for nucleosynthesis up to heavy elements, depending on the temperature at which carbon burning takes place.

I would now like to turn to the *p*-process elements and discuss the possibility of their formation by positron capture during nucleosynthesis. A more complete discussion is given in Reeves and Stewart (1965).

For temperatures of $1\text{–}2 \times 10^{9}$ °K, significant numbers of positrons are formed through pair creation. For example, at 10^{9} °K and a density of 10^{5} gm/cm^{3}, the number of positrons is 10^{25}/cm^{3}; at 2×10^{9} °K and 10^{5} gm/cm^{3}, the density is 2×10^{28}/cm^{3}. Some of these positrons can be captured by s-process nuclei, forming an isotope which can return back to the original nucleus by positron decay, or, part of the time, decay by electron emission into a stable p-element. An example of this is illustrated in Figure 4

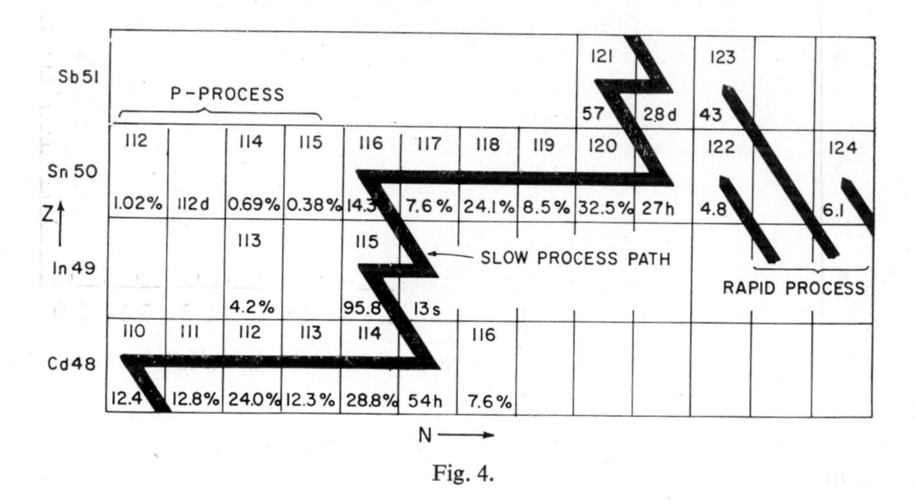

Fig. 4.

(from Clayton, et al.) where the jagged line of the s-process path passes through the region Cd-Sb. If Cd^{114} captures a positron, forming In^{114}, then there is the possibility of In^{114} beta-decaying into the p-process nucleus Sn^{114}.

We have computed lifetimes against positron capture for several s-nuclei in the $1\text{–}2 \times 10^{9}$ °K temperature range. The most promising case was for capture on Dy^{164} leading to the unstable isobar Ho^{164} which can electron decay to Er^{164}, whose large abundance was a difficult point in $B^{2}FH$. For a temperature of 1.5×10^{9} °K and density of 10^{5} gm/cm^{3}, the positron capture lifetime on Dy^{164} is about one hundred years, which is probably comparable to the time that a star spends in the oxygen burning stage. We can then expect some significant abundance of Er^{164} to be built up this way. Since the p abundances are always one or two order of magnitude smaller than the s abundances, we expect that in real life, the positron capture lifetime were smaller than the lifetime in the corresponding stellar stages.

If, on the other hand, the opposite situation had applied, we would have build a state of equilibrium between electron and positron capture and emission processes, hence between the p, s and r representants of a given isobar. The equilibrium abundance between the s and the p element of a given siobar would have been roughly given by the formula $n_s/n_p = e^{\Delta M/kT}$, where M is the mass difference between these two nuclei. In Table IV, the values of n_s/n_p and ΔM are listed.

The absence of correlations between these numbers, and in particular the low abundance of Cd^{108}, Er^{164}, W^{180}, show clearly that the state of equilibrium was not reached in the stellar interiors.

Table IV

Mass number	$\log_{10} n_s/n_p$	ΔM(MeV)
74	2.46	1.206
78	1.80	2.877
84	2.00	1.795
92	2.00	1.658
96	0.65	2.723
102	1.71	1.175
106	1.40	2.779
108	1.53	0.276
112	1.22	1.931
120	2.21	1.700
124	1.66	3.050
126	2.30	0.899
130	1.50	2.549
132	2.34	0.892
136	2.18	2.590
144	1.35	1.817
164	1.44	0.082
168	2.37	1.653
180	1.58	0.165
184	2.41	1.609
190	3.05	1.240
196	2.87	1.667

One could, in principle, use these abundances to put an upper limit to the period of electron processes. At the present time however, it is not possible to calculate the actual value of this limit, since for most nuclei we do not know the energies and the ft values of the excited states.

REFERENCES

Clayton, D. D., Fowler, W. A., Hull, T. E. and Zimmerman, B. (1961) *Ann. Phys.* **12**, 331.

Greenstein, J. and Wallerstein, G. (1964) *Astrophys. J.* **139**, 1163.

Hayashi, C., Hoski, R. and Sugimoto, D. (1962). *Prog. Theor. Phys. Suppl.* 22.

Hoffmeister, E., Kippenhahn, R. and Weigert, A. (1964). *Z. für Astrophysik*, **59**, 215.

Reeves, H., Report, Sept. 1964. *A review of nuclear energy generation in stars, and some aspects of nucleosynthesis*, in "Stellar Evolution" Plenum Press, 1966. p. 83.

Reeves, H. and Stewart, P. (1965). *Astrophys. J.* **141**, 1432.

DISCUSSION

A. G. W. Cameron: One of the outstanding features of the p-nuclei is that their abundances are extremely smooth, whereas anything depending on the results of positron capture should not be expected to be smooth because of the wide distribution of matrix elements. Also, the case you have presented of Er^{164}, I consider as an extreme anomaly if regarded as a p-process nucleus. I have always regarded it as a possible branch of neutron capture taking place slowly. Furthermore, one should include the possibility of excited state electron decay from Dy^{164} to Ho^{164}, because the difference in ground states of the two is only one MeV.

H. Reeves: I would like to make it clear that I do not regard positron capture as the only way to make the p-process elements, but we have studied some twenty cases and the lifetimes against positron capture usually came out to be of the order of the evolution times.

H. Y. Chiu: I might remark that the calculation is very model dependent, in that positron number densities are strong functions of temperature and density.

Current Research at Orsay on the Nuclear Formation Cross-Section of Lithium, Beryllium and Boron

HUBERT REEVES

I would like to talk about some experimental and theoretical work of astrophysical interest that is now being done at Orsay, France, on spallation cross-sections of light nuclei. With the help of refined mass spectroscopic technology (1), they have been the first to measure the cross-section for the formation of the stable lithium isotopes (Li^6 and Li^7) by bombardment of high energy protons on various targets. They have also set up a program (2) for statistical calculations of these cross-sections. The calculated values are in good agreement with experiments.

The principle of these calculations is a Monte-Carlo analysis of the intra-nuclear cascade initiated by the incoming protons, followed by a phase space calculation of the breaking-up of the excited nucleus left at the end of the cascade. They assume that the break-up probability of this residual nucleus is proportional to the phase space allowed when you consider various spallation possibilities.

The same technique has been applied to calculate the cross-sections for the formation of various isotopes of beryllium and boron. Some of the results are presented in Table I.

I have listed the Be^{10}/B^{10} ratio mostly because it turns out to be so small. Be^{10} has been used as a measure of irradiation time-scale, as its decay would influence the B^{11}/B^{10} ratio. The Orsay result implies that this may be delicate, since so little Be^{10} seems to be formed.

Using these results, together with many others, on the spallation of light nuclei, the isotopic ratios of lithium and boron formed by the bombardment of stellar gases of ordinary cosmic mixtures of elements with protons of typical "cosmic ray spectrum" have been evaluated (3).

The ratios at ($t = \infty$) are $n_7/n_6 = 2.5 \pm 1$; $n_{11}/n_{10} = 5 \pm 2$; they are only mildly sensitive to the shape of the spectrum or to the exact composition of the gas as long as these are kept within "realistic" limits. The uncertainty attached to the ratios reflects this sensitivity.

Table I

Isotopic formation ratios from proton bombardment of C^{12} and O^{16} at various energies (1), (2) and (3). The ratios labelled ($t = 0$) are measured immediately after the nuclear reactions; the ratios labelled ($t \to \infty$) are measured after all β unstable products have had time to decay. The values given in column 3 and 4 come partly from experiments done in other laboratories, partly from calculations done at Orsay.

	Energy (MeV)	Li^7/Li^6 ($t \to \infty$)	B^{11}/B^{10} ($t \to \infty$)	Be^{10}/B^{10} ($t = 0$)
$p - C^{12}$	50	3.6	—	—
	150	2.0	5	0.1
	550	2.2	4	0.1
$p - O^{16}$	150	1.9	2.0	0.1
	600	$\leqq 1.9$	2.0	0.1
	Meteoritic	12.5	4	—
	Stars	$2 \to \infty$	—	—

The difference between the n_7/n_6 formation ratios and the stellar observed ratios is usually assigned to (p, α) reaction taking place at the bottom of the surface convective zone of a star where Li^6 burns much faster than Li^7; this is consistent with the fact that no star has been observed where the ratios are smaller than the formation ratios.

For the case of the meteorites, the observed boron ratio is, within experimental uncertainties, the same as the formation ratio in typical cosmic gases.

This suggests that the planetary light elements (Li, Be, B) may have been formed on the surface of the sun, where the lithium isotopic ratio could be altered by (p, α) reaction, but where the boron isotopic ratio (because of the increased Coulomb repulsion), would remain unaltered.

This model, an alternative to the F. G. H. model (4) of neutron irradiation of cold planitesimals, is developed in (2) and (3).

Note added in proof: Measurements of Be^9, Be^{10}, B^{10}, B^{11} (5) essentially corroborate the conclusions presented here.

REFERENCES

(1) Bernas, R., Epherre, M., Gradsztajn, E., Klapisch, R. et Yiou, F. (1965). *Phys. Letters* **15**, 147.

(2) Gradsztajn, E. (1965). Thése Orsay, *Ann. de Physique* **10**, 791.

(3) Bernas, R., Gradsztajn, E., Reeves, H., Schatzmann, E. "On the nucleosynthesis of lithium, beryllium and boron". To appear in *Annals of Physics,* Oct. 1967.

(4) Fowler, W. A., Greenstein, J. I., and Hoyle, F. (1962). *Geophys. J., Roy. Astro. Soc.* **6**, 148.

(5) Yiou, F., Baril, M., Dufaure de Citres, J., Fontes, P., Gradsztajn, E., Bernas, R. Submitted to *Physical Review.*

Stellar Evolution through Helium Burning*

Icko Iben

This paper is a brief description of salient agreements and disagreements between the theory of stellar evolution and observations. Some attention will be given to the detailed evolutionary history of a single three solar mass star, with comments on the evolution of different masses between 1 and 15.

Observational evidence for the birth of stars is ubiquitous, of course. It exists in the form of the luminous O and B stars which have lifetimes known to be much less than the age of the earth. More direct evidence, in the sense of observing something today which wasn't there yesterday, is more difficult to find. Perhaps a classical example of this rarer evidence is FU Orionis which brightened by a factor of a hundred in luminosity in a matter of about 120 days some thirty years ago. Things such as M-82 might be taken as evidence for the birth of massive Fowler-Hoyle type objects.

The theory for birth of a star out of gas and dust is extremely rudimentary, and begs the question as to how the gas and dust got there in the first place. Theory concerning the behaviour of a condensation once it reaches a density such that the mean free path for a photon is small compared to the dimensions of the system, is on a much firmer foundation. Hayashi[1] has shown that, for stars of a given mass, there exists a forbidden region in the Hertzsprung-Russell diagram, in which models in static equilibrium cannot be constructed. There are a number of theoretical uncertainties concerning the exact position of this boundary. The relationship between surface temperature and luminosity is highly sensitive to the type of opacity chosen in the narrow photospheric layer. Most of the star is convective in this portion of its evolution, but convective flow of energy at optical depths of one or less is not properly understood. By merely varying the parameters in the only theory[2] available, this boundary may be moved considerably. Perhaps a more fundamental criticism of stellar models in this region of the H-R diagram is the fact that in the photosphere, where all the radiative transfer

* not revised since delivery, January 1965.

[1] See Hayashi (1962), p. 81–85, for example.

[2] The mixing length theory is presented in Vitense (1953).

occurs, the mean free path for a photon is comparable or larger than the temperature scale height, which violates the diffusion approximation so commonly used in stellar evolution calculations.

The behaviour of FU Orionis is evidence for this instability region being one of dynamic motions. Its track in the H-R diagram might look like that shown in Figure 1, a collapse from a radius of perhaps 100 $R_\odot$ to 20 $R_\odot$ in a

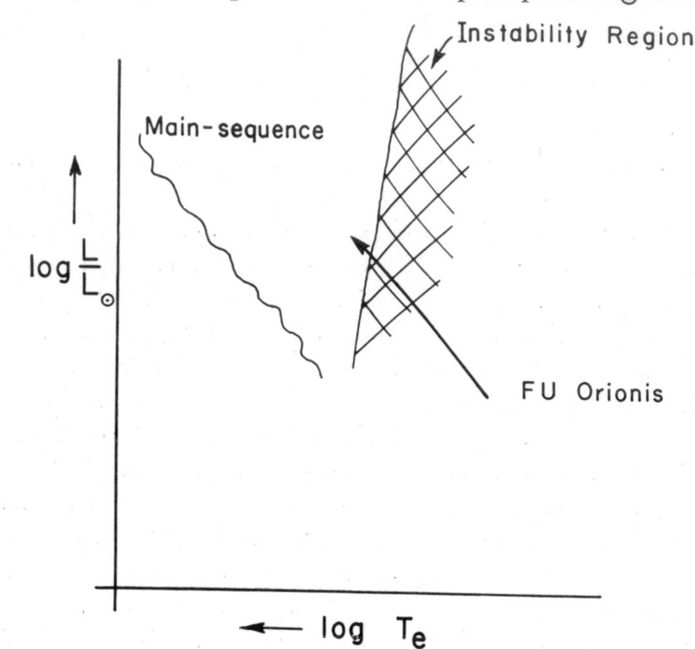

Fig. 1. A rough sketch of the region of dynamic instability in the Hertzsprung-Russell diagram. The instability region encompasses all portions in the diagram to the right of and including the cross hatched region bounded by a Hayashi evolutionary track.

matter of 120 days. The fact that FU Orionis stayed in this region for a considerable time (at least it was observed there before 1938) indicates that there are probably other regions in the H-R diagram in which quasi-stable models can be made. Construction of such models requires a more complete understanding of other sources of opacity in the low temperature—low density region, due to molecular hydrogen, water vapor and so on. I think that the case of FU Orionis shows that the lifetime that has previously been placed on gravitational contraction may be an underestimation. The existence of regions of stability to the right of the Hayashi track in the H-R diagram (see Figure 1) would allow the star to evolve on a time scale longer than the presently-assumed hydrodynamic time scale through this region.

Evidence for the validity of the Hayashi track is the position of the T-Tauri stars in the H-R diagram. Hayashi (1961) has shown that the T-Tauri stars in the cluster NGC-2264 seem to fall along the stable side of the boundary. Loci for constant time of a number of evolutionary tracks for stars of different mass, assuming no mass loss, can be fitted to observations reasonably well. However, there is considerable mass loss during the vertical descent along the Hayashi track, with perhaps a 40% loss from a $1M_{\odot}$ star. Thus, the apparent agreement between time-constant loci, for models without mass loss, and observation must be considered with some care.

The ramifications of this Hayashi phase for nucleosynthesis are manifold; only a few will be discussed. First, the theory of the production of light elements such as Li, Be and B by spallation of high energy protons and heavier elements, as proposed by Fowler, Greenstein and Hoyle (1962) encounters difficulties. The temperature of the solar nebula is higher than they assumed because the luminosity is larger Also, the time scale is now a bit shorter than was expected before Hayashi's analysis. Kuhi's (1964) study of mass loss from T-Tauri stars indicates that the flux of high energy protons is higher than might be desired for this model.

If there are other regions in which a star may spend time, the difficulty with the time scale is removed. Also, FU Orionis was variable for a long period of time before 1938, indicating the possible occurrence of large flares which would cause the proper spallation reactions. The observation of lithium in FU Orionis and in T-Tauri stars suggest that these stars, which are presumably contracting, have made lithium in surface reactions. Thus the fact that on the main sequence the lithium content of a star increases as the mass increases, may be explained in terms of the details of Li-burning in the convective envelope during the contraction phase. A correlation between age on the main sequence and Li abundances indicates that part of the burning occurs on the main sequence.

It is well known that there is general agreement between the observations and theory for the main sequence in the H-R-Diagram. Constructing models of uniform composition, various age-zero main sequences may be obtained by varying the heavy element content, keeping hydrogen fixed, or vice versa. In the lower portion of the main sequence there is large uncertainty due to inadequate knowledge of convective flow. Consequently, comparison of theory with observation is extremely difficult in this region. On the upper main sequence the slope predicted by almost any opacity is nearly identical with that obtained from observation of young clusters.

In the mass-luminosity plane, a set of curves can be constructed for stellar models having homogeneous composition; these are shown in Figure 2. With increasing hydrogen content and fixed heavy element abundance Z,

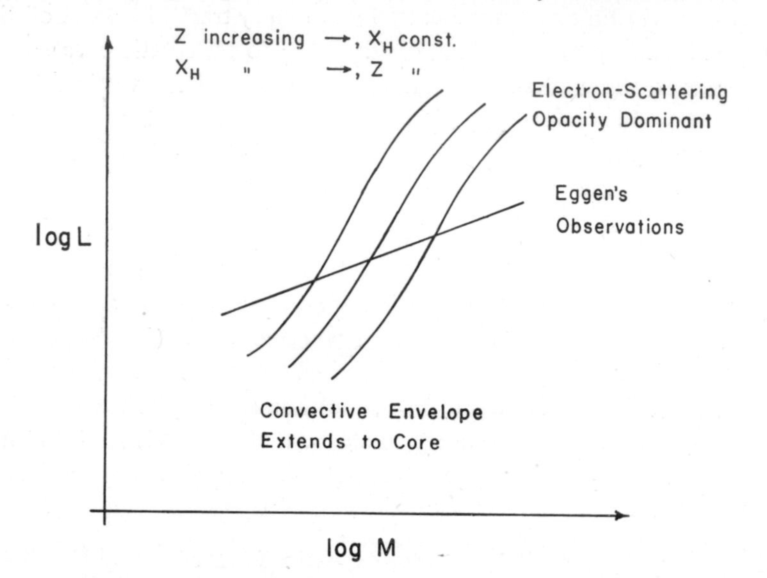

Fig. 2. A sketch of theoretical initial main sequences in the mass-luminosity diagram, showing the dependence of position on the composition choice (X = abundance by mass of hydrogen and Z = abundance by mass of heavy elements). Eggen's empirical mass-luminosity relationship for the Hyades-Pleiades "group" has a much shallower slope than that given by theory.

progress is made to the next curve to the right in the diagram. In the same way, with fixed hydrogen content, increasing Z causes motion to the right. In the low luminosity region of the curves, the convective envelope extends to the inner portion of the star, while in the high luminosity region the opacity due to electron scattering dominates the energy flow in the star.

If the nearby stars of Population I composition are plotted in the mass-luminosity diagram, they fall along the theoretical tracks, even down to low masses where the curve turns. On the upper portion (high masses) there is some divergence from the theoretical curve which cannot be accounted for on the basis of evolution.

A great discrepancy does exist however. If the results of Eggen's (1963) empirical mass-luminosity determination of the Sirius group, the Hyades, and the Pleiades are plotted, as in figure 2, the curve is observed to have a slope much less than predicted by theory. This discrepancy can not be removed by reasonable variations in opacity. The observations are of well-known groups which have been investigated often. There are two ways out of this difficulty. One is that the mechanism of formation of these stars is such that more massive stars favor high hydrogen relative to helium content, and less massive stars are formed with more helium and less hydrogen. That is, the composition is somehow dependent upon the mass. This is hard to believe, but it may possibly be a constraint that any theory of stellar formation must accommodate.

Another possibility of escape from the difficulty is rotation. Preliminary work by Roxburgh (1964) and others indicates that luminosity tends to decrease for a given mass as one increases the angular velocity of the object. In any case, if the models are correct a discrepancy seems to exist which hasn't been considered thoroughly.

Evolution off the main sequence will be sketched briefly. Figure 3 shows the general behaviour of evolutionary models of higher mass as they begin to leave the main sequence. This behavior agrees well with that observed in some clusters, so that the theory appears satisfactory. For stars of lower mass, the evolutionary tracks vary with the mass of the star. Figure 4 shows roughly the behavior of a 1.0 $M_{\odot}$ and 1.5 $M_{\odot}$ star.

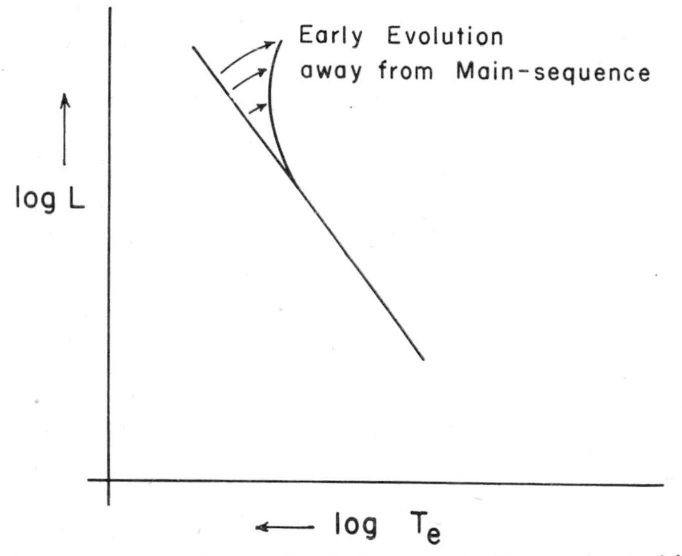

Fig. 3. Arrowed lines are sketches of evolutionary tracks (for massive stars) joining an initial main sequence with a time constant locus in the H-R diagram. Stars on a time constant locus are all of the same age.

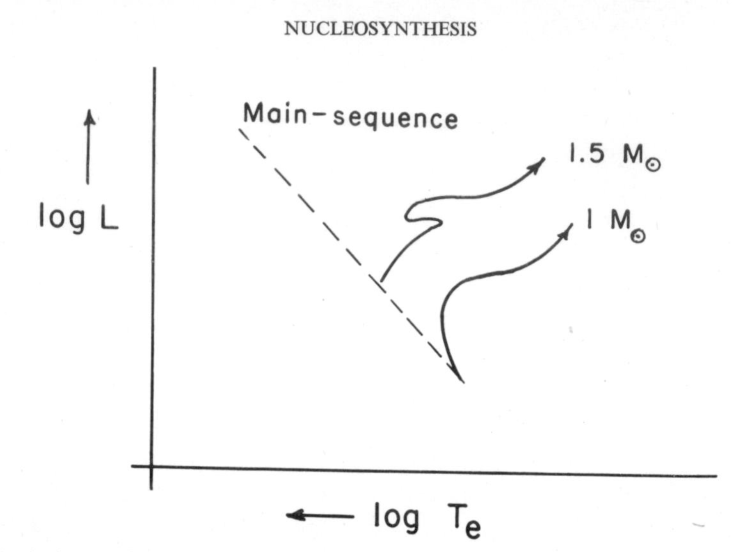

Fig. 4. A sketch of evolutionary tracks in the H-R diagram for low-mass stars. The hook along the 1.5 $M_\odot$ track occurs as hydrogen is exhausted over a large central region once in a turbulent-convective state. The 1 $M_\odot$ star possesses no convective core at any stage and hydrogen is progressively exhausted over only small portions of the energy producing regions.

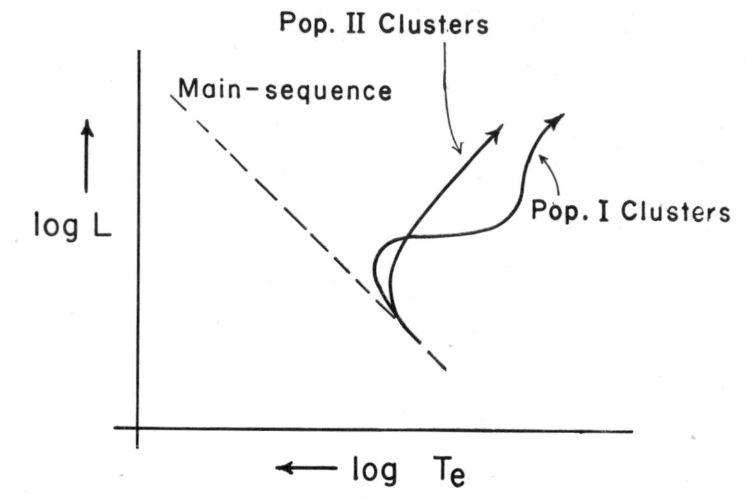

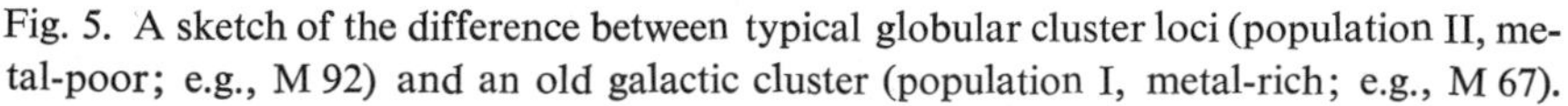

Fig. 5. A sketch of the difference between typical globular cluster loci (population II, metal-poor; e.g., M 92) and an old galactic cluster (population I, metal-rich; e.g., M 67).

In the H-R diagram, Population I clusters (such as M-67 and NGC-188) and Population II clusters fall on curves something like those shown in Figure 5. There is much better agreement between loci of constant time for theoretical models for various masses and the observed shapes of cluster diagrams for Population I stars. At present there is some discrepancy in shape between theoretical tracks and cluster diagrams for Population II stars. I think that this difficulty with the shapes is simply a result of incomplete treatment of energy transfer, both by photons (inadequate opacities) and by convection which has considerable influence in this low temperature region.

There is another discrepancy, better-known, between the ages determined for old clusters and from the Hubble constant. A number of suggestions have been made to alleviate this discrepancy, one being that of Clayton (1964): The ages of clusters as determined by fitting with theoretical time-constant curves, may be lowered by allowing for some mass loss of order of 10^{-11} $M_{\odot}$ per year. Another suggestion is that of Dicke (1962), in which the universal gravitational constant, G, changes with time. Still another possibility is to suppose that Population II stars have high helium contents. Although this is contrary to the ideas of some researchers, my feeling is that high helium content is at the root of the problem. This may come about as follows: suppose that initially there is a galaxy of gas and dust with no heavy elements. If condensations of the order of 10^5 to 10^7 solar masses could occur, theory predicts that they would be dynamically unstable upon reaching hydrogen burning, and completely convective. Thus much hydrogen may be processed to helium, and a lot of hydrogen and helium thrown off, without increasing much the abundance of heavy elements.

There is another point of agreement between observation and stellar models. Figure 6 shows the evolutionary track of a 3 $M_{\odot}$ star which has a convective core during its hydrogen-burning phase on the main sequence. On the portion of the track between 3 and 4 rapid gravitational contraction occurs when hydrogen abundance drops to nominal values in the center of the star. Hydrogen-burning begins in a shell outside this hydrogen-exhausted core as the star evolves from 4 to 6. Along 6 to 10 the core contracts and the envelope expands as the shell source continues to burn. At the luminosity minimum at 10, a convective envelope develops, becoming larger as the luminosity rises to the red giant "tip" at 13 where helium-burning by the three-alpha process begins in the core. The main phase of core helium-burning occurs from 14 to 18, with central helium exhaustion at 20. The stage of hydrogen-burning in a shell source, especially the path from 4 to 6 requires a considerable time interval, as compared to the gravitational time scale. This portion of the evolution and the phase of core helium-burning (that is, the path from 14 to 18) require roughly one fourth of the total evolutionary time from the onset of hydrogen-burning at 1, to central helium exhaustion at 20.

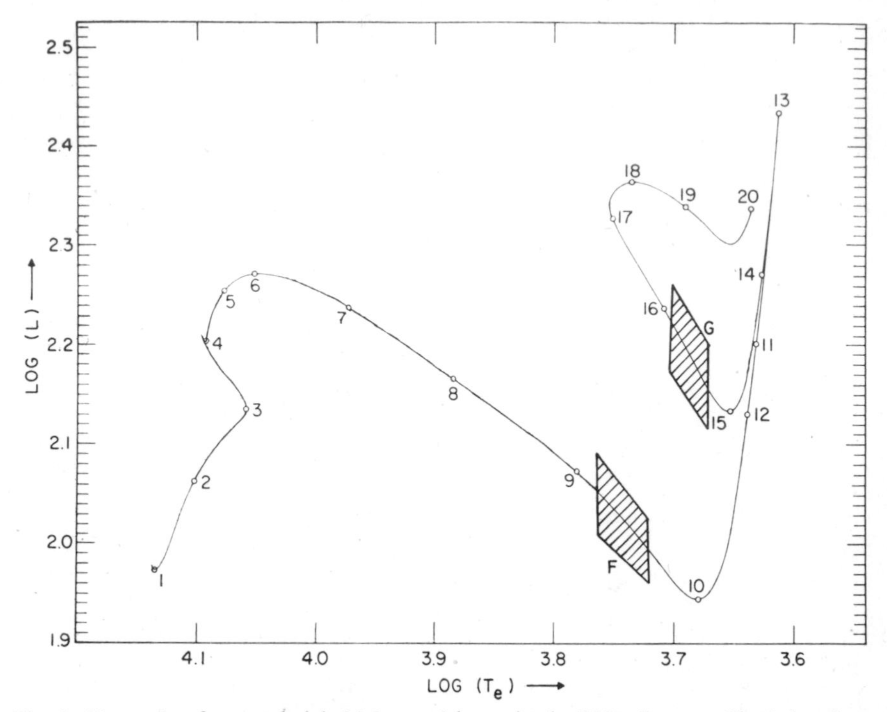

Fig. 6. The path of a metal-rich 3 $M_\odot$ model star in the H-R diagram. Shaded regions represent positions of two components of Capella, component G having very little surface Li and component F having a surface abundance of Li roughly 100 times the solar value.

The time spent in hydrogen-burning in a shell source relative to gravitational contraction time goes up as one increased the mass of the star. By considering an appropriate cluster, gaps should appear in the H-R diagram where the theoretical models show that evolution proceeds rapidly, and groups of stars should lie in regions where evolution proceeds slowly. That this is observed in M 67 is another point of agreement of theory and observation.

Helium-burning stages may naturally be discussed in two groups, the distinction being whether or not electron degeneracy occurs in the core at the onset of helium-burning. For Population I, a star with mass greater than 2.5 $M_\odot$ will be nondegenerate in its core. By constructing curves like Figure 6 for a number of different masses (see Figure 7), it is found that helium-burning in the core and hydrogen-burning in a shell occurs in a rather wide region parallel to the hydrogen-burning main sequence, as shown in Figure 8. By constructing stellar envelopes to discover in what positions in the H-R diagram pulsational instability occurs, a region is found like that shown

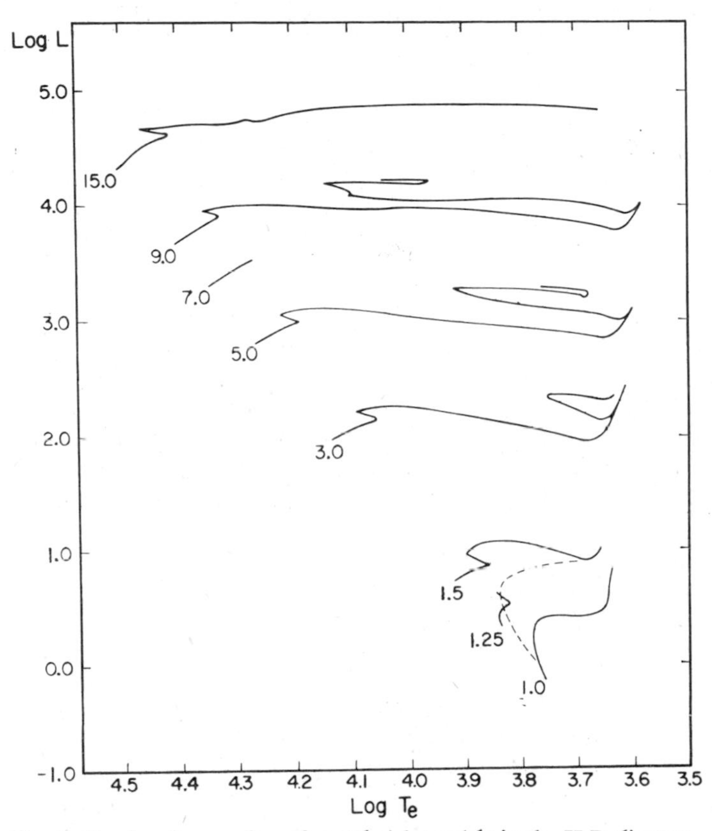

Fig. 7. Tracks of a number of metal-rich models in the H-R diagram.

in Figure 8. In the region in which pulsational instability, and in which He-burning in a core with an H-burning shell are predicted, relatively long-lived, pulsating stars are expected. In fact, this is the region in which Cepheid variables are found. It appears, therefore, that Cepheid variables correspond to the theoretical models.

Still another point of contact between theory and observation lies in the surface abundances of light elements. As the envelope becomes more and more extended, hydrogen and helium become partially ionized over a large portion of the atmosphere, and conditions become favorable for convective energy transport. As the star gets larger and larger, the convective region extends further and further, including a greater mass of the star. Light elements such as Li, Be, and B would perish in the interior during the main sequence phase. This large outer convection zone would mix the interior regions depleted in these light elements, so that the surface abundance of these elements would change. This period of evolution, in which there is a deep convective envelope, is comparatively rapid (see Figure 6, path 10 to 13). Thus one expects to see different surface abundances for stars of the same initial abundance, depending upon whether they are in the slow-time scale region before or after the onset of helium burning.

Figure 6 has two shaded regions which correspond to the observed positions in the H-R diagram of the two components of Capella. A calculation of the destruction of lithium during the main sequence phase indicates that Li is destroyed over 99% of the interior. When the convective envelope extends almost to the hydrogen-burning shell during the later evolution, the surface abundance of Li drops by a factor of 80. Observations indicate that Capella F does have an overabundance of Li with respect to Capella G of about two orders of magnitude. This indicates that mixing does occur in this stage of evolution.

At the end of helium-burning in the core, the photoneutrino process must be considered. As the core contracts to the point at which carbon-burning or oxygen-burning can occur the pair annihilation process becomes important. Caution is urged when considering models in which these processes have not been included.

There are perhaps two major uncertainties associated with helium-burning. Because of the uncertain nature of the convective transport parameters, the position of the heliumburning region may be varied with some latitude. A second difficulty has to do with nucleosynthesis. Because of uncertainty in the reduced width of the reaction C^{12} (α, γ) O^{16}, it is uncertain as to whether the product of helium-burning will be oxygen, carbon, or some intermediate mixture of the two. As regards the comparison with observations of models undergoing He-burning in their cores, the uncertainty makes little difference because the hydrogen-burning shell supplies most of the star's luminosity. He-burning in the core simply stabilizes the shell temperature; the amount of energy coming from the helium core relative to the amount of energy from the H-burning shell is about 20% over most of the He-burning phase.

For stars less massive than 2.5 $M_{\odot}$, most effort in model building for the He-burning phase has been concerned with Population II stars. Only Harm and Schwarzschild (1964) have examined what happens during the helium flash period in any detail. However, the radius of the star is uncertain when the center evolves violently. Plasma neutrino losses were neglected, although according to Chiu (1963) this mechanism may account for 10% to 40% of the star's energy loss. Also, although time steps were of the order of seconds, dynamic terms were neglected in the equations of motion.

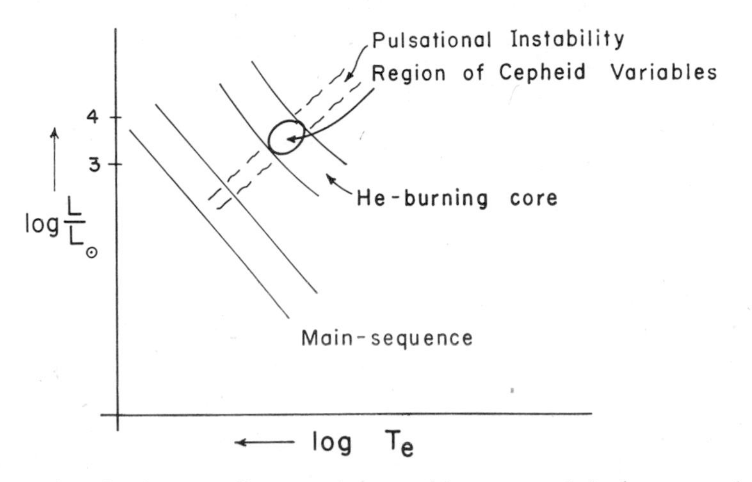

Fig. 8. Regions in the H-R diagram where model stars spend the largest portions of their active nuclear-burning lives. On the main sequence, stars are burning hydrogen at their centers. Stars burning helium at their centers and hydrogen in a shell define an additional "main-sequence" displaced above the hydrogen-burning main-sequence.

All other work concerned with He-burning phases for low mass stars has consisted of guessing as to what happens after the helium flash, the guess being constrained by an attempt to fit the stars on the horizontal branch in the H-R diagram. The mass in the form of helium after the flash phenomena is varied; this should be greater than or equal to the mass in the helium core before the flash. The total mass of the star, after the flash may also be varied, assuming some mass loss occurred. The amount of C^{12} and O^{16} formed during the flash might also be varied.

Hayashi (1962), Nishida and Sugimoto (1962), and others have shown that by choosing parameters properly, a horizontal branch may be reproduced in the H-R diagram. The time scale presents problems; it appears that there isn't enough time during this phase to account for the density of stars in the horizontal branch. I feel that instead of guessing, a tremendous effort should be put into examining what happens at the helium flash stage, including dynamic terms.

What happens inside a star? Figure 9 shows the behavior of several variables as the $3M_\odot$ star evolves. L is luminosity in solar units (3.86×10^{33} erg/sec), T_e the surface temperature, M_{cc} the mass fraction in the convective core, and X_i the abundance by mass at the stellar center of the most important nuclide of atomic weight i (for example, X_4 refers to He^4, X_{12} to C^{12}, X_{18} to O^{18}, and so on). The hydrogen burning phase occupies approximately the first

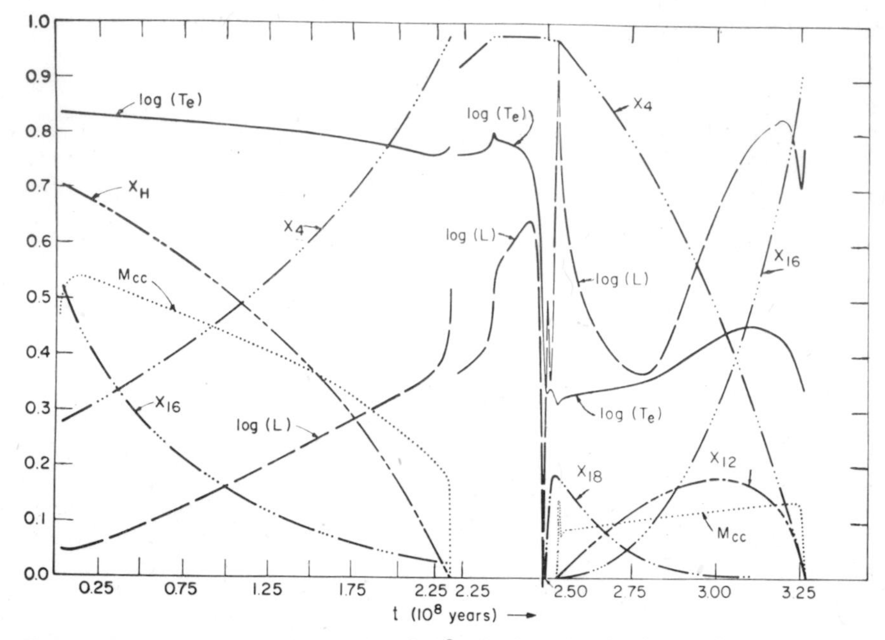

Fig. 9. The variation with time (units of 10^8 yr) of luminosity (L), surface temperature (T_e), mass fraction in the convective core (M_{cc}), and central abundance by mass of $H^1(X_H)$, $He^4(X_4)$, $C^{12}(X_{12})$, $O^{16}(X_{16})$, and $O^{18}(X_{18})$. The unit of luminosity is $L_\odot = 3.86 \times 10^{33}$ erg/sec and the unit of surface temperature is °K. Vertical scale limits correspond to $1.95 \leq \log L \leq 2.45$, $3.3 \leq \log T_e \leq 4.3$, $0 \leq M_{CC} \leq 1/3$. To the left of the berak in t, $0.0 \leq X_{16} \leq 0.02$, $0.0 \leq X_H$, $X_4 \leq 1.0$. To the right of the break in t, $0.0 \leq X_{18} \leq 0.1$ and $0.0 \leq X_4$, X_{12}, $X_{16} \leq 1.0$.

2.52×10^8 years. During this phase the luminosity steadily increases. After this the luminosity has another maximum associated with H-burning in a thick shell source. Then the star rapidly evolves to the red giant region, core He-burning starts, and the abundance of He^4 begins to decrease at the center. The luminosity again begins to increase as the star settles down to He-burning in the core with an H-burning shell source.

The fine structure in the luminosity curve at the red giant peak ($t \sim 2.50 \times 10^8$ years) is due to the exhaustion of N^{14} by $N^{14}(\alpha, \gamma)O^{18}$, which briefly reverses the evolution, sending the luminosity lower. Now Ne^{22} has been suggested as a possible source of neutrons. In Figure 9 the abundance of O^{18} is shown decreasing after ($t \sim 2.50 \times 10^8$ years) due to the formation of Ne^{22}. However the maximum temperature at the end of helium-burning is only 1.6×10^7 °K which is not high enough for neutron production. There are no alpha particles left for the $Ne^{22}(\alpha, n)Mg^{25}$ reaction by the time the temperature is high enough. These results are for a 3 $M_\odot$ model, but it is generally true that He-burning occurs at about 1.0×10^8 °K rather than the 2.0×10^8 °K necessary to allow neutron production by Ne^{22} (α, n) Mg^{25}. In order for this neutron production mechanism to work in the shell, energy loss by neutrinos must force the shell source to higher temperatures to supply more of the energy.

REFERENCES

Chiu, H. Y. (1963). *Astrophys. J.* **137**, 333.
Clayton, D. D. (1964). *Astrophys. J.* **140**, 1064.
Dicke, R. H. (1962), *Rev. Mod. Phys.* **34**, 110.
Eggen, O. J. (1963). *Astrophys, J. Suppl.* VIII, No. 76, 125.
Fowler, W. A., Greenstein, J. L. and Hoyle, F. (1962). *Geophys. J.* **6**, 148.
Harm, R. and Schwarzschild, M. (1964). *Astrophys. J.* **139**, 594.
Hayashi, C. R. Hoshi, and Sugimoto, D. (1962). *Prog. Theor. Phys. Suppl.* **22.**
Nishida, M. and Sugimoto, D. (1962). *Prog. Theor. Phys.* **27**, 145.
Roxburgh, I. W. (1964). *M.N.R.A.S.* **128**, 157; **128**, 237.
Vitense, E. (1953). *Z. für Astrophys.* **32**, 135.

DISCUSSION

W. Bidelman: I have just two comments. There is a difference in rotational development of the two components in the H-R diagram. The one on the left side is rotating more rapidly than the one on the right. It looks as if rotation is slowed during the passage.

Second, you must be careful about placing stars in H-R diagrams, expecially in the case of FU Orionis. In the theoretical H-R diagram, absolute magnitudes are used. At the present moment, FU Orionis has effectively no bolometric corrections, so the observed magnitude is an absolute magnitude. At an earlier stage it may have been much cooler. If so the mere process of heating would make it appear to brighten. It is not impossible that instead of following the path in the H-R diagram that you suggested, it's path was almost horizontal. The point I wish to make is that the star did not necessarily change its *absolute* luminosity by six magnitudes although its *visual* luminosity may have changed quite a bit.

G. Wallerstein: We have learned with velocity relations in the Hyades that we can set some limits on the possible composition dependence of the stars along the main sequence. Conti, Wing and I have determined hydrogen to metal abundance ratios of stars from class AO to K2. The hydrogen-to-metal ratio is constant, so you can't vary the helium abundance much. I don't think that it is a free parameter.

W. L. W. Sargent: Talking with Eggen, I got the impression that for widely separated pairs he used an average to get the massluminosity relation. This has the effect of reducing the slope in the center of the mass-luminosity curve, relative to theoretical values.

B. G. Stromgren: Taking the mass-luminosity law from what the observers consider to be reliably-determined binary pairs, this small set of data, analyzed in terms of He/H ratio, gives results which are remarkably constant over a wide range of masses. If less certain cases, such as Eggen included, are used difficulties in theoretically fitting the He/H ratio appear. The real question is whether part of these difficulties might not be due to systematic errors. It is really a difficult observational problem. I think there is no doubt that even within Population I stars, there is a certain range of variation in the helium-hydrogen abundance ratio.

W. L. W. Sargent: If I may add something to the last remark, the mass-luminosity relation is very well applied to binary pair with large parallaxes, talking *well-determined* binaries from Van de Kamp's article[3] in Handbuch de Physik. Going to smaller parallaxes, significant deviations in both directions are seen in what appears to be the mass-luminosity relation.

[3] P. van de Kamp (1958). "Visual Binaries", Handb. der Phys., Springer-Verlag, ed. S. Flügge, vol. L, 187.

Evolution of a Star of 30 Solar Masses

R. STOTHERS

The evolution of a 30 $M_\odot$ star is representative of very massive stars. The reason for choosing the particular value of 30 $M_\odot$ is simply that this mass has been featured in discussions of Type II supernova explosions (e.g., Hoyle and Fowler 1960; Fowler and Hoyle 1964). The initial composition (by mass) is assumed to be: hydrogen, $X_e = 0.70$; helium, $Y_e = 0.27$; and heavy elements $Z_e = 0.03$. The carbon-nitrogen-oxygen abundance is $X_{CNO} = Z_e/2$.

A main-sequence star of this mass lies so far on the early side of the H-R diagram that its surface temperature is quite high; consequently the opacity is assumed to be due entirely to electron scattering. The equation of state is represented by the sum of perfect-gas and radiation pressure. The effect of semiconvection (see below) is fully taken into account.

Mass loss is assumed to be negligible on the basis of the available observational data. Miss Underhill (1960) obtained mass loss rates equal to 10^{-7} $M_\odot$/year for O and B stars, running perhaps as high as 10^{-6} $M_\odot$/year for the Wolf-Rayet stars. The time scale for significant mass loss from a 30 $M_\odot$ star is then

$$\tau \sim \frac{M}{(-dM/dt)} = 10^7 \text{ to } 10^8 \text{ years},$$

which is much longer than the evolutionary time scale.

In this paper, the results of a calculated sequence of stellar models for 30 $M_\odot$ (Stothers 1963, 1964, 1966) will be summarized. We wish to focus particular attention on the gross structural and nuclear evolution of the star.

Figure 1 shows the distribution of hydrogen as a function of mass fraction for the stellar models during hydrogen burning. Hydrogen is depleted in a shrinking convective core in a period of about five million years. Outside the convective core there is a small region of convective instability. In this

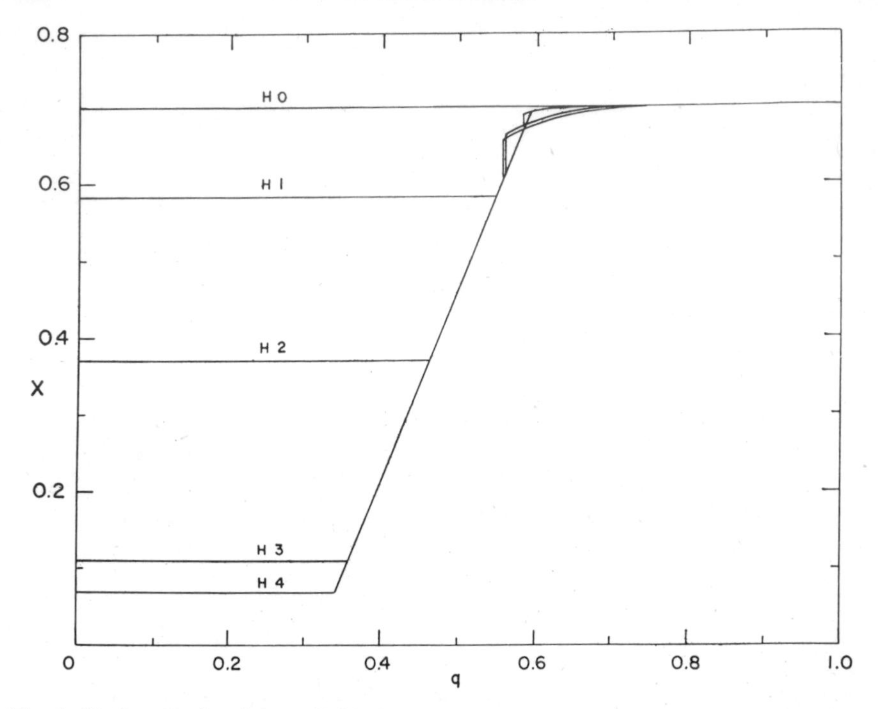

Fig. 1. Hydrogen abundance, X, is plotted against mass fraction, q, for five stages of hydrogen burning (H).

region, mass motions slowly and continually readjust the gradients of hydrogen and helium to preserve a quasi-static situation where convective neutrality is approximately maintained. This gives rise to the phenomenon that has been called "semiconvection". The net result is to mix the helium-enriched interior composition outward into the envelope while at the same time hydrogen in the outer region is mixed inward. The long diagonal curve in Figure 1 represents the distribution of chemical composition that would be characteristic of the intermediate zone if there were no instability outside the convective core, that is, if the region were assumed to be in complete radiative equilibrium. The semiconvective mixing, by bringing hydrogen down into the interior, modifies this distribution somewhat, and raises a segment of the curve as shown.

At the interface with the semiconvective zone, the radiative intermediate zone shows a jump in chemical composition. In reality, there will be a very steep gradient in mean molecular weight here, but we have approximated it by a discontinuity.

Figure 2 represents the evolution of the interior zones as a function of time, for the phases of hydrogen burning, hydrogen exhaustion, and gravitational contraction in the core. Notice that the semiconvective zone grows monotonically outward into the radiative outer envelope of age-zero composition, and at a faster rate than it grows inward. As hydrogen is ex-

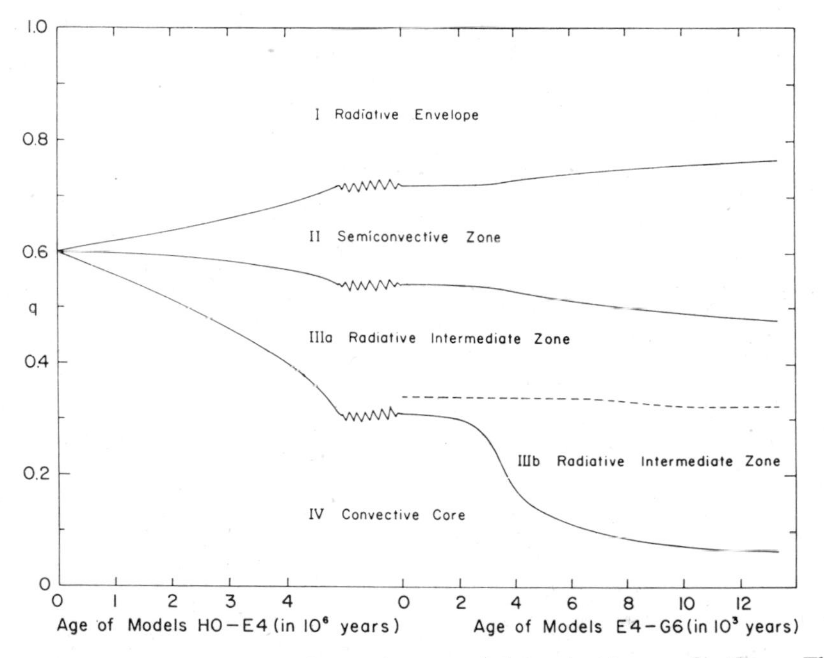

Fig. 2. Evolution of the structural zones in terms of their extent in mass fraction, q. The phases included are: (1) initial main sequence (H0) to the hydrogen-exhaustion stage where $X_c = 0.001$ (E4), and (2) the latter stage (E4) to the end of core gravitational contraction (G6). The dashed line represents the hydrogen-burning shell.

hausted near the center, the mass fraction of the convective core shrinks rapidly, but never vanishes because the radiation pressure remains high relative to the gas pressure. This maintains the convective instability. The dotted line represents the hydrogen shell source, which starts burning as soon as the contraction of the core has raised the shell temperature sufficiently.

Figure 3 shows the evolutionary track of the star in the H–R diagram. As the core burns its hydrogen, the star leaves the zero-age main sequence and evolves upward and to the right (points *a* to *b*). When the central hydrogen content is nearly exhausted, gravitational contraction of the whole star

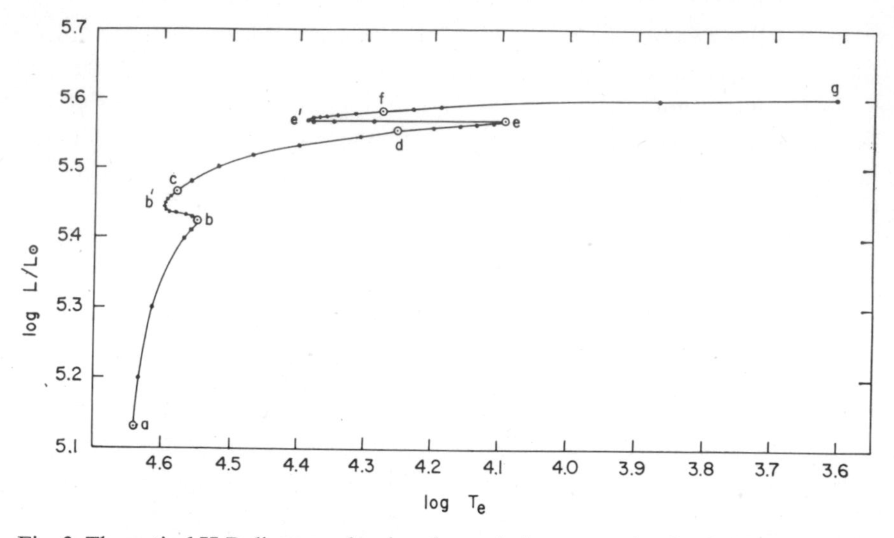

Fig. 3. Theoretical H-R diagram, showing the evolutionary track of a 30 M⊙ star. The phases are as follows: *a-b*, hydrogen burning; *b-c*, hydrogen exhaustion; *c-d*, core gravitational contraction; *d–e*, helium ignition; *e–f*, helium depletion; *f–g*, helium exhaustion.

sets in. The evolutionary track of the star then turns to the left, toward smaller radii. When the shell is ignited, the envelope begins to expand again (point *b'*).

From point *c* on, gravitational contraction of the core proceeds to raise the central temperature until helium burning commences. At the same time the luminosity of the hydrogen shell source increases, bringing the star over to point *d* in the H–R diagram. Actually the contribution of the shell source to the total luminosity is not dominant; it averages only 50% during this phase. Unlike the situation in lower mass stars, a great portion of the luminosity of a 30 M⊙ star is due to gravitational contraction.

At point *d*, helium is ignited in the central region. The shell is still contributing appreciably to the luminosity, however, so that the radius continues to expand until helium burning contributes the dominant portion of the luminosity (point *e*). From the commencement of helium burning in the core until helium has been depleted by about 70%, the luminosity remains remarkably constant, changing by less than 1%. The total radius shrinks, however, as the mass fraction of the shell moves significantly outward. After the helium content in the core drops below 0.3 (point *e'*), the position of the shell is stabilized on account of the short time scale, and the evolutionary track turns back to the right. The luminosity begins to increase more noticeably. This trend continues until helium is practically exhausted in the core, and once again gravitational contraction sets in (point *f*). During the phase of central helium burning the major portion of the luminosity comes from the helium-burning core, with very little contributed by the hydrogen-burning shell. At the onset of helium exhaustion, the luminosity contribution from the shell is only 7%. Near point *g*, photoneutrino emission becomes important, and at this point the calculations were terminated.

Figure 4 represents the structural evolution of the star during the advanced phases of helium ignition, helium depletion, and helium exhaustion. These phases last slightly more than half a million years. It is clear that the semiconvective zone does not change much during helium ignition and helium

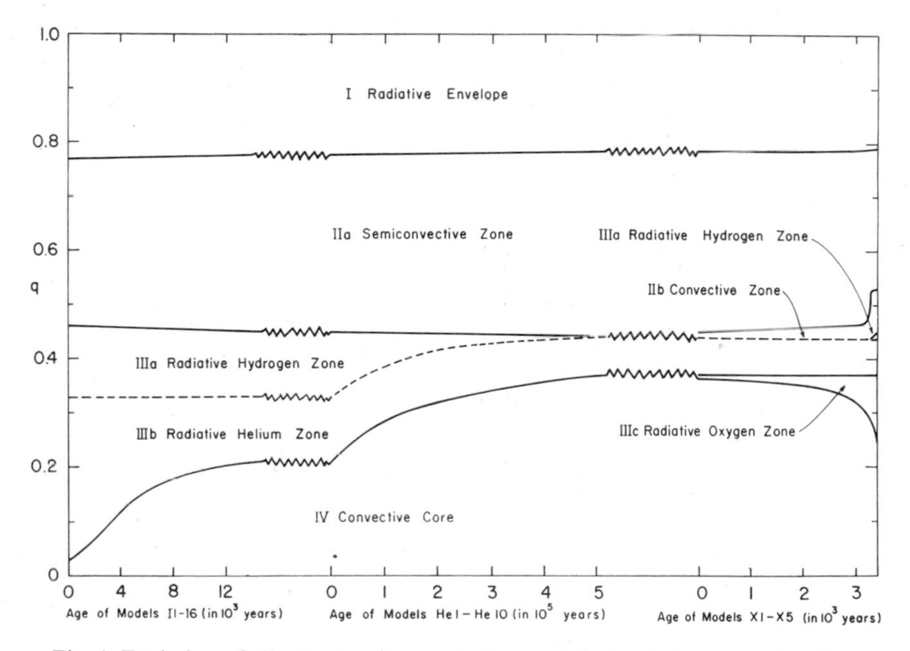

Fig. 4. Evolution of the structural zones in terms of their extent in mass fraction, *q*. The phases included are: (1) helium ignition (I1–I6), (2) helium depletion (He1-He10), and (3) helium exhaustion (X1-X5). The dashed line represents the hydrogen-burning shell.

burning because the total luminosity remains almost constant. The hydrogen-burning shell gradually consumes its way toward the surface, because the time scale of evolution during helium burning is much longer than during the preceding gravitational contraction phase. The mass fraction of the convective core increases rapidly as helium is consumed, nearly catching up with the hydrogen-burning shell. Ultimately, the shell reaches the semi-convective zone. At this point an abrupt change occurs in the hydrogen abundance. However, the shell luminosity is already so low that the sudden 50% increase in hydrogen has very little effect upon the structure of the model.

Peculiar things happen as the star enters the helium exhaustion phase. The major point to note is that the inner boundary of the semiconvective zone starts moving out toward the surface, while a fully convective zone develops between it and the radiative helium zone. Many treatments of this region were attempted to determine the structure of the star during this phase; the present result appears to be the only self-consistent solution. The complication arising from instability in the envelope is the main reason why computations of very massive stars have been avoided in the past. The models presented here are at least self-consistent; whether semiconvection or something else actually occurs in a star is another matter.

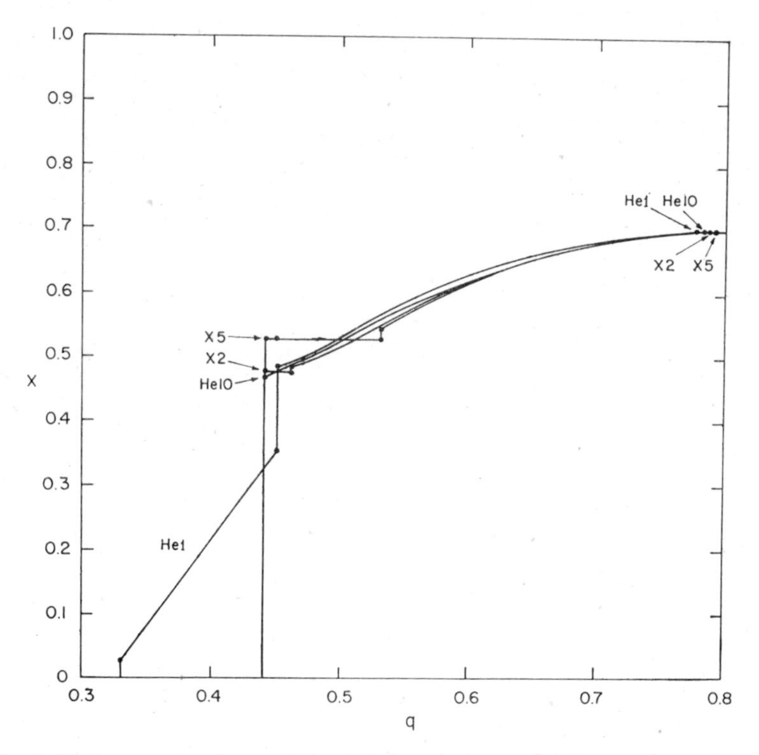

Fig. 5. Hydrogen abundance, X, is plotted against mass fraction, q, for various stages of helium burning (He) and helium exhaustion (X).

Figure 5 represents the hydrogen distribution in the star during the helium-burning phases. The shrinkage of the semiconvective zone becomes apparent, as it is pushed from beneath by the small convective zone. The interface between the two zones ought to be, strictly speaking, a very thin radiative region with a steep composition gradient. However, for the sake of simplicity we have assumed a composition discontinuity for the interface. At first, the fully convective zone increases its extent outward to about 10% in mass fraction, while its inner boundary remains nearly fixed. After a while, like the semiconvective zone, the fully convective zone also moves as a whole toward the surface. The effect of the unstable zones is to maintain a net downward transport of hydrogen, while at the same time to mix helium up toward the surface.

During the period of development of the fully convective zone, the convective core shrinks as helium is exhausted, leaving behind a region which will probably be composed mainly of oxygen. The exact composition depends upon the rate of the reaction $C^{12}(\alpha, \gamma)\,O^{16}$, and therefore upon the undetermined value of the α-particle reduced width, Θ_α^2, for the O^{16} compound nucleus.

Now that the general aspects of the evolution of a 30 $M_\odot$ star have been presented, the evolutionary changes in the chemical composition will be discussed in more detail. The initial composition of the star was

$$X_e = 0.70$$
$$Y_e = 0.27$$
$$Z_e = 0.03$$

During hydrogen burning, the mean chemical abundances are altered to

$$\bar{X} = 0.38$$
$$\bar{Y} = 0.59$$
$$\bar{Z} = 0.03$$

Almost half of the initial hydrogen content is consumed. The gravitational contraction phase changes these abundances very little. However, after helium burning, the new abundances are

$$\bar{X} = 0.36$$
$$\bar{Y} = 0.25$$
$$\bar{Z} = 0.39$$

Very massive stars, starting with a solar-type initial composition on the main sequence, have less helium in their final stages than they started with! Notice also that, once the main phase of core hydrogen burning is finished, very little further depletion of hydrogen takes place in the shell. At helium exhaustion, the core abundances of carbon and oxygen are (neglecting the possible synthesis of heavier elements)

$$X_C = 0.29$$

$$X_O = 0.71$$

for a reduced width $\Theta_\alpha^2 = 0.1$. If $\Theta_\alpha^2 \geqq 0.3$, the composition will be entirely oxygen.

The star may now be considered a presupernova. However, the calculated structure will be only formal since mass loss and rotation were neglected. According to the evidence in O associations of the Large Magellanic Cloud, the Wolf-Rayet phase seems to occur near the end of hydrogen burning (Westerlund 1961; Westerlund and Smith 1964; Stothers 1965). Thus the relevant evolutionary time scale may be only of the order of 10^5 years, making the mass loss estimate for hydrogen burning even smaller than before However, during helium burning when the luminosity and radiation pressure are much higher and a shell source is ignited, pulsational instability might set in, relieving the star of considerable mass (cf. Schwarzschild and Härm 1959). Mixing between the deep convective envelope and the interior zones when the star becomes a red supergiant at the onset of carbon or oxygen burning might even produce explosive mass loss. Finally, rotational mixing during any stage of the evolution could cause marked changes in the interior structure.

The work reported here was supported by an NAS-NRC research associateship under the National Aeronautics and Space Administration. It is a pleasure to thank Dr. Robert Jastrow for his hospitality at the Institute for Space Studies.

REFERENCES

Fowler, W. A., and Hoyle, F. (1964). *Astrophys. J. Suppl.* **9**, 201.
Hoyle, F. and Fowler, W. A. (1960). *Astrophys. J.* **132**, 565.
Schwarzschild, M. and Härm, R. (1959). *Astrophys. J.* **129**, 637.
Stothers, R. (1963). *Astrophys. J.* **138**, 1074.
(1964). *Astrophys. J.* **140**, 510.
(1965). *Astrophys. J.* **141**, 671.
(1966). *Astrophys. J.* **143**, 91.

Underhill, A. B. 1960. *Stellar Atmospheres*, ed. J. Greenstein (Chicago: University of Chicago Press), p. 411.
Westerlund, B. (1961). *Uppsala Ann.* **5**, No. 1.
Westerlund, B. E. and Smith, L. F. (1964). *Mon. Not. R.A.S.* **128**, 311.

DISCUSSION

Dr. Iben: I notice that both you and Hayashi[1] find that at the onset of the triple-α reaction your models move toward higher surface temperatures at relatively constant luminosity. Hayashi did not include semiconvection in his calculations, nor the possible absorption or release of gravitational potential energy in the envelope. I do not get this constancy of luminosity for my 15 $M_\odot$ star. Why?

Dr. Stothers: I did not include the gravitational energy involved in the contraction and expansion of the envelope either. I think that during helium burning the time scale is slow enough that this effect should be small. As to the question of semiconvection, I re-integrated one of Hayashi's envelopes for the helium-burningphase of a 15.6 $M_\odot$ star and found that what I described had already happ ened—the semiconvection seems to disappear as the luminosity gets higher. During helium burning he had effectively no semiconvection left. What would remain is either a radiative zone or a very small fully convective zone. This shouldn't have much effect. I don't know what caused the difference you mention, unless it is the opacity or our assumption that the shell is negligibly thick.

Dr. Iben: As the mass increases between 3 $M_\odot$ to 9 $M_\odot$, the contraction after the onset of the triple-α is relatively rapid, and gravitational energy contributes heavily to the luminosity.

Dr. Stothers: Do you mean in the envelope?

Dr. Iben: Yes.

Dr. Stothers: The cases of 15.6 $M_\odot$ and 30 $M_\odot$ are then very different from 9 $M_\odot$

Dr. Reeves: If I understand correctly, in your model during helium burning most of the energy comes from the core and not the shell, while I think Dr. Iben has the opposite result. Is this a question of mass?

Dr. Stothers: Yes.

[1] See S. Sakashita, Y. Ono, and C. Hayashi, (1959). Prog. Theoret. Phys. (Kyoto) **21**, 315; and C. Hayashi and R. C. Cameron, (1962), Astrophys. J. **136**, 166. (Ed. note).

The Evolutionary State of Peculiar A Stars[1,2]

John Faulkner

After submitting the title for my contribution, it became clear that a slightly lengthier title would have been preferable, namely "The Evolutionary State of the Theory and Observations of Peculiar A Stars."

In a paper about to be submitted to the Astrophysical Journal, Fowler, Burbidge, Burbidge and Hoyle (1965; hereinafter FB^2H) take the view that the particular combination of abundance peculiarities found in the Ap's can only be formed at a late stage of stellar evolution, after one, or a series of helium-flashes. As my own work has recently been concerned with models for post-helium-flash stars, it is of interest to see whether explicit models can lend support to this suggestion.

At this point I should perhaps answer Iben's criticism of those who investigate later stages of evolution prior to conquering all the preceding difficulties. Carried to its logical conclusion this argument would eliminate all stellar model investigators (not necessarily a bad idea), as no one has yet satisfactorily demonstrated how a star can form. It is my own belief that a suitable combination of observation and theory can help one determine how a star picks its way through the difficulties in practice. The whole object, therefore, of investigating the current models is to see whether, as we Englishmen would say, our ideas about the helium-flash are, in fact, on the right croquet lawn. My conclusion (Faulkner 1966) is that they are.

Eggen (1959) came to the conclusion that the Ap's formed a coherent linear feature in the color-magnitude diagram. The line (with appreciable scatter) had a somewhat shallower slope than the main sequence, and was brighter than the main-sequence by $\sim 0\overset{\mathrm{m}}{.}7$ to $\sim 1\overset{\mathrm{m}}{.}0$ in the range $0.0 \lesssim B\text{–}V$

[1] Supported in part by the Office of Naval Research [Nonr-220(47)] and the National Science Foundation [GP-5391].

[2] A talk delivered at the NASA Conference on Nucleosynthesis, New York, 25–26 January 1965.

$\lesssim 0.2$. For $B—V \lesssim 0.0$ there was an indication that the Ap's turned upwards to keep essentially parallel to the main sequence. This and other information is summarized in Figure 1 of FB²H.

It is difficult to see how even pure hydrogen main-sequence stars could exist this high in the CM diagram, with such a disparate type of slope. Of course, the spectral peculiarities might for some reason necessarily manifest themselves only in a certain range of surface parameters, and one just happens to have caught stars passing through this region in the normal course of their evolution away from the main sequence. This seems, however, an artificial solution to the problem.

I wondered whether there was any other type of structure which could account for a reasonably long-lived, high luminosity, shallow-sloped feature. It turns out that there is. A post-helium-flash star with an ignited helium core is very sensitive to the abundance of CNO elements adjacent to the core (Faulkner 1966). If the CNO group is absent, the hydrogen shell-source relies only on the pp-chain. This permits higher densities and temperatures in the shell and envelope. The radius of the whole star is then considerably smaller than it would be with C, N and O present. Stars with a given core mass $\sim 0.5\ M_\odot$, and varying total mass, $1.5 \gtrsim M/M_\odot \gtrsim 1.0$ do indeed define the type of feature noticed by Eggen.

Thus, by a curious coincidence, the structure which appears to be necessary requires the CNO group to be essentially absent, in agreement with FB²H. Furthermore, as mass (and luminosity) of the models increases, the pp-shell source temperature becomes so high that around $\log T_e \gtrsim 4.0$ a NeNa cycle can begin to operate, and the sequence of stars will change their slope, as Eggen observed.

There are, however, some arguments and recent observations which can be set against this picture. If the stars have passed through the helium flash, they must have originally had $M/M_\odot \lesssim 2.5$—yet Kraft has found Ap's in clusters with a main-sequence up to ~B0. Similar objections hold for the two Ap's in M 39. Baschek and Oke (1965) suggest that care should be taken before assigning positions in the CM diagram for the Am's and Ap's. The Am's which appear to fall on the end of Eggen's Ap relation have considerable errors in the colors owing to crowding of the lines of the many rare earths. Indeed, the blanketing may amount to as much as $\sim 0\overset{m}{.}2$ in $B—V$. This is quite alarming, for when corrected, the stars fall to the left of the main sequence. This being the case there is really no need to look for an exotic structure to explain the positions of such stars. We simply have something we think we understand anyway. However, if the line-blanketing results are confirmed in the sense that corrected Ap stars lie below the main-sequence, we are still in some difficulties. Helium is observed to be of low abundance in Ap's yet we know that main-sequence stars with low helium should lie along the upper edge of any main sequence band. This suggests that helium cannot be deficient throughout the stars, and that the deficiency must be merely a surface effect. Confirmation indeed for the pre-1965 FB^2H theory!

This seems a fitting note on which to end. We have heard considerable discussion about the abundances of elements in certain stars from study of their spectra. A remark made by Lyttleton many years ago may well prove relevant in this regard. Lyttleton stated that "If one believes that one can determine the distribution of the elements inside stars from their surface composition, then one may as well believe a chimney sweep is made of solid carbon."

REFERENCES

Baschek, B. and Oke, J. B. (1965) *Astrophys. J.*, **141**, 1404.

Eggen, O. J. (1959). *Observatory*, **79**, 197.

Faulkner, J. (1966) *Astrophys., J.*, **144**, 978.

Fowler, W. A., Burbidge, E. M., Burbidge, G. R. and Hoyle, F. (1965), *Astrophys. J.*, **142**, 423.

The Approach to Supernova Implosion Conditions

H. Y. CHIU

I would like to start by discussing under what conditions we can expect neutrino emission processes to dominate the luminosity of a highly evolved star.

The relaxation time τ_{ph} for a star to be cooled by photon emission is given roughly by

$$\tau_{ph} = \frac{R^2}{C\lambda}$$

where R is the stellar radius and λ is the photon jean free path. From the virial theorem and a reasonable value for λ this yields

$$\tau_{ph} = 2 \times 10^{11}\ \mathrm{T}_9 \text{ sec.}$$

The corresponding time via pair annihilation neutrino emission τ_ν is

$$\tau_\nu = 1.5 \times 10^4 \exp(11.9/\mathrm{T}_9)\, \mathrm{T}_9(\mathrm{M}_\odot/\mathrm{M})^2, \ \mathrm{T}_9 \ll 6.$$

At $\mathrm{T}_9 = 1$, $\tau_\nu = 2 \times 10^7$ sec and $\tau_{ph} = 2 \times 10^{11}$ sec.

We conclude, that for $T_9 \gtrsim 1$ photon energy transport cannot be important in the center of stars. Thus, the march of evolution for these highly evolved stars can be calculated using neutrino luminosities only.

If photon effects are ignored ("neutrino star") then the stellar structure equations can be written

$$\frac{dP}{dr} = -\varrho \frac{GM_r}{r^2}$$

$$\frac{dM_r}{dr} = 4\pi r^2 \varrho$$

with an energy generation rate

$$\varepsilon = \varepsilon_{gr} + \varepsilon_n + (dv/dt) = 0$$

where ε_{gr} is gravitational and ε_n is nuclear. We may set $\varepsilon_{gr} = -T(dS/dt)_M$. [See Chiu (1964)].

Is has been proposed by some authors that solutions of these equations can be obtained by assuming a polytropic form for the relation between pressure and density, viz.,

$$\varrho = K\varrho^{1+\frac{1}{n}}$$

and then solving the resultant Lane-Emden equation in the static ase. If the polytropic index is held constant in time then the evolutionary progress of the star can be found using homology relations. The result obtained is that $\varrho^{\alpha}T^3$. An implicit assumption of this type of calculation is that energy can be redistributed inside the star at a much faster rate than it can be produced by gravitational contraction. Fowler and Hoyle (1960, 1964) have used this scheme, which works very well for some classes of star, to evolve very late stars. They conclude that a star of mass around 30 $M_\odot$ will collapse when the temperature reaches $\sim 6 \times 10^9$ °K and a density of 2×10^7 gm/cm³. In what follows I shall take issue with this result and indicate that collapse will take place long before *via* neutrino processes.

The first point to consider is whether homologous contraction can be applied to neutrino stars. Consider the case where all nuclear processes are turned off, i.e. set ε_n equal to zero. Then

$$T(dS/dt)_M = -(\varepsilon_{\nu 0}/\varrho)\, T^k.$$

For nondegenerate stars, where $T_9 \gtrsim 3$, $\varepsilon_{\nu 0} = 4{\cdot}3 \times 10^{15}$ and $K = 9$. For a nondegenerate gas, S is given by

$$S = R_g\, l_n(T^{\alpha}\varrho^{-1}) + \text{constant},$$

where α is a slowly varying constant equal to 1·5 for nonrelativistic matter and equal to 3 for relativistic. For $T_9 = 2$, $\alpha \approx 2{\cdot}7$. If we assume a relation between T and ϱ of the form

$$(\varrho/\varrho_0) = (T/T_0)^{\beta}$$

where ϱ_0, T_0 and β are constants, then we have

$$(dT_9/dt) = \gamma T_9^{\,K-\beta}$$

where

$$\gamma = \frac{\varepsilon_{\nu 0}\, T_0^{\beta}}{\varrho_0 R_g(\beta - \alpha)}$$

This can be solved to yield

$$T_9 = (T_9^0)/[1 - (t/t_0)]^{(1/(K-\beta-1))}$$

where

$$t_0 = \left(\frac{1}{K - \beta - 1}\right)\left(\frac{\beta - \alpha}{\alpha}\right)\tau_\nu.$$

For a homologously contracting neutrino star, collapse occurs when $t = t_0$. If we use $\beta = 3$, $K = 9$, $\alpha = 2.7$, then $t_0 = (1/45)\,\tau_{\nu'}$, and free fall for a 10 $M_\odot$ star is reached when $T_9 \sim 3$.

The collapse can be delayed to higher temperatures if β is close to $K - 1 = 8$. An exact calculation, done by myself and E. E. Salpeter indicates that β is close to 8 for the central regions of the star. For 10 $M_\odot$ and polytropic index $n = 1.5$ with a good equation of state including radiation, electrons and positrons and heavy nuclei, the $\varrho - T$ line for the star is given initially by a straight line of slope 1.5. If the star evolves as a polytrope then this line will be displaced without distortion along a line of slope 3 (Figure 1). The actual evolution is very complicated as shown in Figure 2. Since plasma neutrinos dominate at high densities, the loss rate increases with density and a positive temperature gradiant ensues in the inner third of the stellar mass and the polytropic index, instead of being $n = 3$, is rather $6 \leqq n \leqq 9$. Thus, the use of polytropes is inadequate.

If we go on to include the nuclear energy sources from $O^{16} — O^{16}$ and $C^{12} — O^{16}$ burning, then gravitational contraction in a region of burning can take place only along curves given by

$$\varepsilon_{gr} + \varepsilon_n + (dU/dt) = 0.$$

Since $\varepsilon_{gr'}$, ϱ and T appear in logarithms, the condition $\left(\frac{dU}{dT}\right) + \varepsilon_n \simeq$ constant $\times\ T$ determines the temperature and density relation according to which the star can evolve. We find that for conditions of interest ε_n is given by

$$\varepsilon_{16-16} = 8 \times 10^{-5} \varrho T_9{}^{33} \text{ erg/gm-sec},$$

$$\varepsilon_{12-16} = 1.32 \times 10^2 \varrho T_9{}^{31.8} \text{ erg/gm-sec}.$$

Thus, the $\varrho - T$ paths for evolution are

$$8 \times 10^{-5} \varrho\ T_9{}^{33} + (4.3 \times 10^{15}/\varrho)\, T_9{}^{33} = \text{constant } T$$

for $O^{16} - O^{16}$, and

$$1.32 \times 10^2 \varrho\ T_9^{31.8} + (4.8 \times 10^{18}/\varrho)\, T_9{}^3 \exp\,[-2(T_0/T_9)] = \text{constant } T$$

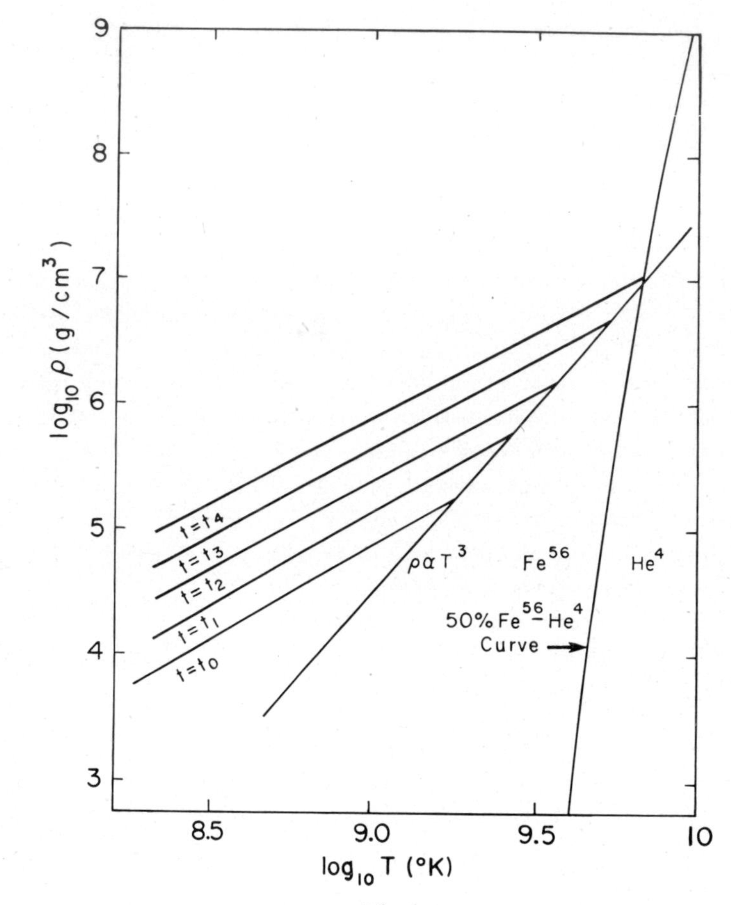

9
8
7
6
5
4
3
$\log_{10} \rho$ (g/cm^3)
8.5
9.0
9.5
10
$\log_{10}$ T (°K)
t=t4
t=t3
t=t2
t=t1
t=to
ρ α T 3
Fe56
He4
50% Fe56 – He4
Curve

Fig. 1.

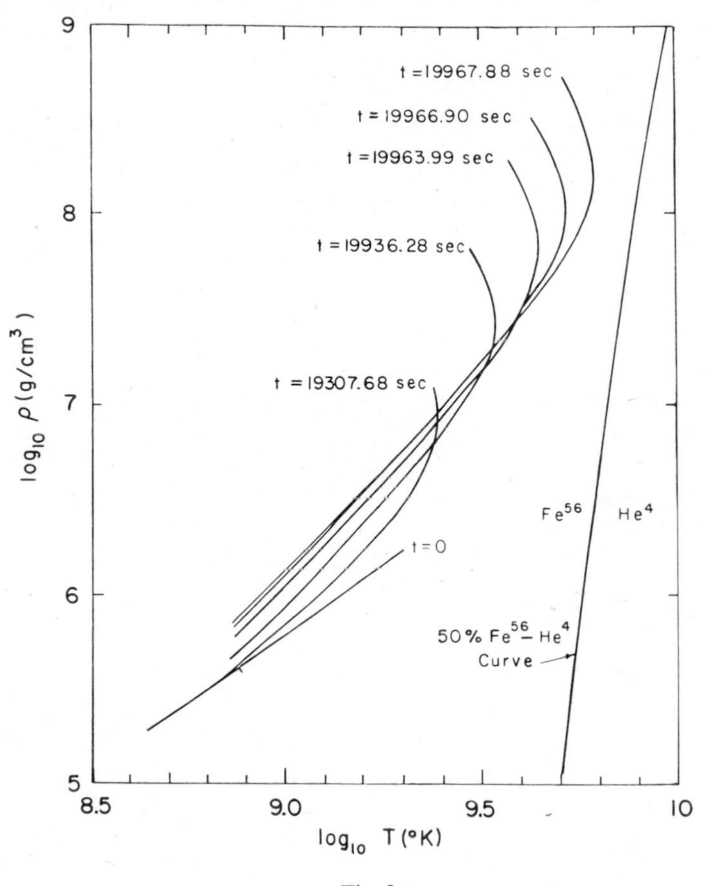

Fig. 2.

for $C^{12} - O^{16}$ burning. In the $\log \varrho - \log T$ plane these are lines with negative slope which we shall call $n - \nu$ curves. The particular curves with the constant replaced by zero will be called the $n - \nu$-zero curves shown in Figure 3. The quanties x_a, in the Figure are the concentrations of element a.

A neutrino star in the nuclear burning stage is expected to follow the $n - \nu$-zero curves very closely because of the very sharp temperature dependence of the nuclear energy rates. If the star is on the low temperature side the gravitational contraction will increase the temperature until burning commences and stops the contraction and the star settles on the $n - \nu$-zero line. Too high an initial temperature causes expansion and the temperature drops until a balance is reached, and so on.

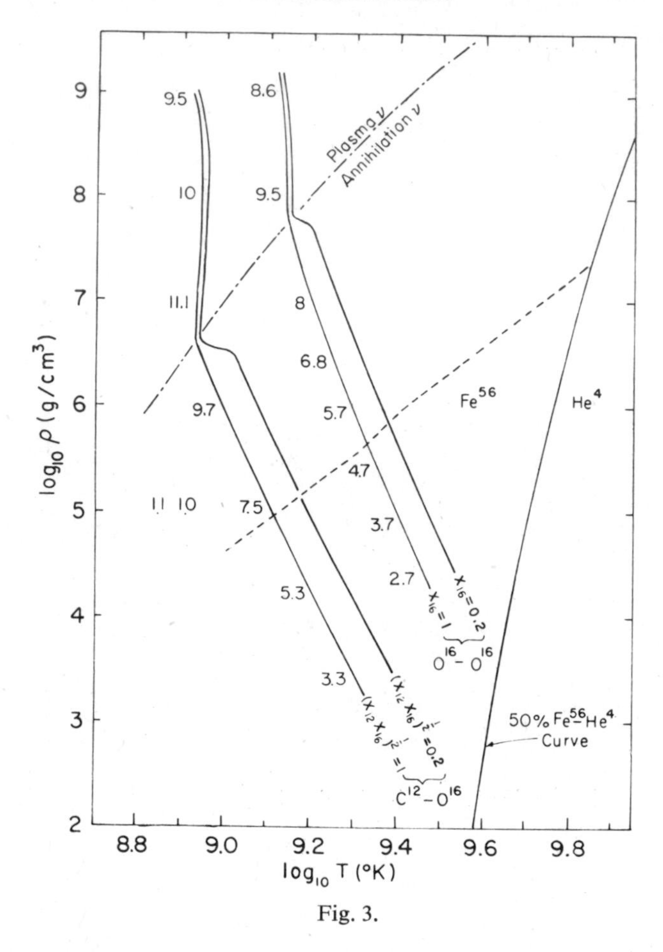

Fig. 3.

When the density reaches 4×10^7 gm/cm^3 for $O^{16} - O^{16}$ burning or 4×10^6 gm/cm^3 for $C^{12} - O^{16}$ burning, then plasma neutrinos become important and, as shown in Figure 3, the $n - \nu$-zero line approaches the vertical and the temperature becomes independent of density and is a constant.

The time scales for evolution also differ drastically from those predicted on the basis of simple polytropes. As indicated in Figure 3, the star evolves quickly (in 10^7 sec) at a temperature of around 1.2×10^9 °K ($C^{12} - O^{16}$) or 2.5×10^9 °K ($O^{16} - O^{16}$), its density increasing rapidly. The time scale then becomes longer ($\sim 10^8 - 10^9$ sec) for $O^{16} - O^{16}$ burning and $\sim 10^{11}$ sec for $C^{12} - O^{16}$. It is not known if the density will reach a value of $\sim 10^{10}$ gm/cm^3.

Fowler and Hoyle have argued that the transition from iron to helium at a density of around 10^7 gm/cm^3 and a temperature of 7×10^9 °K is sufficient to cause collapse in a 30 $M_\odot$ star. However, it seems that this density and temperature cannot be reached simultaneously in neutrino stars. It is possible that the final collapse state will occur at very high density ($\sim 3 - 4 \times 10^9$°K), independent of the mass of the star. The mechanism for collapse may e the approach to equilibrium (at very high density) *via* β-transitions or pynonuclear reactions.

REFERENCES

Chiu, H.-Y. (1964). *Ann. Phys.* **26**, 364.
Fowler, W. A. and Hoyle, F., (1960). *Astrophys. J.* **132**, 565.
Fowler, W. A. and Hoyle, F. *Astrophys. J. Suppl.* No. 91.

Statistical Properties of Nuclei

A. G. W. CAMERON, A. GILBERT, C. J. HANSEN, and J. W. TRURAN

We have been concerned with the problem of the synthesis of heavy elements in stellar interiors. In particular, we have sought to follow in detail the nuclear transformations involved both in the approach to nuclear statistical equilibrium and in the production of heavy elements by neutron capture on a fast time scale. These investigations have led, necessarily, to a rather extensive study of nuclear systematics.

The details of stellar evolution through the hydrogen and helium burning phases are quite well established (Hayashi *et al.* 1962). The products of helium burning, C^{12} and O^{16}, will be destroyed at temperatures $T \lesssim 2 \times 10^9$ °K by heavy ion reactions. The details of these reactions are not well determined. The reaction products should be nuclei with masses in the range $20 \lesssim A \lesssim 32$, particularly the alpha-particle nuclei Mg^{24}, Si^{28}, and S^{32}. Although the relative abundances of these products are uncertain, it is clear that as the temperature is increased Si^{28}, having the highest separation energies for protons and alpha-particles, will be the last nucleus to be photodisintegrated and will therefore accumulate. It can be assumed that the medium consists mainly of Si^{28} at temperatures $T \gtrsim 2 \times 10^9$ °K.

In order to follow in detail the transformation of this silicon region to nuclei in the vicinity of the iron peak, it is necessary to obtain thermonuclear reaction rates for all proton, neutron, and alpha-particle reactions proceeding on the intermediate nuclei. As experimental determinations of the resonance parameters are not available in most cases, we must proceed on the basis of the nuclear statistical properties. In particular, this demands a knowledge of the nuclear level densities, the nuclear radiation widths, and the particle widths for protons, neutrons, and alpha-particles. These subjects will be considered in some detail in our subsequent discussions.

A detailed investigation of neutron capture on a fast time scale demands a knowledge of the neutron capture cross sections for nuclei far on the neutron rich side of the valley of beta-stability. For these nuclei, experimental determinations of the neutron binding energies and the nuclear level densities are not available. It will be necessary, therefore, to predict these features by means of a semi-empirical mass formula.

THERMONUCLEAR REACTION RATES

The number of reactions per unit volume per second, r, of two nuclear species with number densities n_1 and n_2 can be written in the form:

$$r = n_1 n_2 \langle \sigma v \rangle$$

Here $\langle \sigma v \rangle$ is an appropriate average of the product of the reaction cross section, $\sigma(v)$, and the relative velocity of the nuclei, v,

$$\langle \sigma v \rangle = \frac{\int \sigma \langle v \rangle \, v N(v) \, dv}{\int N(v) \, dv}$$

where $N(v)$ is the number density of nuclei having relative velocity v.

The determination of the rates $\langle \sigma v \rangle$ as a function of temperature was considered in detail in another paper by the authors (Truran *et al.* 1966). Assuming that the velocity distributions of the two species are Maxwellian, that the contribution to the cross section of a single narrow resonance is given by the Breit-Wigner single level formula, and that there are many resonances in the energy range of interest, the total rate for a particle-particle reaction involving a as the incoming particle and b as the outgoing particle (leaving the residual nucleus in a definite state) is:

$$\langle \sigma v \rangle_{a,b} = \frac{2.51 \times 10^{-13}}{(\mu T_9)^{3/2}} \frac{1}{(2S_a + 1)(2I + 1)}$$

$$\sum_{J,\pi} (2J + 1) \int_{E_t}^{\infty} dE \, e^{-11.61E/T_9} \, \varrho(E, J, \pi) \frac{\Gamma_a \Gamma_b}{\Gamma} \text{ cm}^3 \text{ sec}^{-1}. \qquad (1)$$

In this expression, E_t is the threshold energy for the reaction, μ is the reduced mass in amu, T_9 is the temperature in unit of 10^9 °K, Γ_a and Γ_b are the particle widths in MeV, Γ is the total width for the decay of a specified compound nuclear state (taken to be the sum of the proton, neutron, and alpha-particle widths and the nuclear radiation width), S_a is the spin of the incoming particle, I is the spin of the target nucleus, J and π are the spin and parity, respectively, of the compound nuclear state, and $\varrho(E, J, \pi)$ is the level density of states of specified spin and parity at excitation E. The particle widths, Γ_a and Γ_b, contain summations over all values of orbital angular momentum and channel spin consistent with angular momentum and parity conservation. The corresponding expression for the rate for particle capture reactions is given by equation (1) with $\Gamma_b \rightarrow \Gamma_\gamma$ (Γ_γ being the radiation width for the specified compound nuclear state (E, J, π)).

We must now consider the approximations employed in the determination of the nuclear parameters contained in equation (1).

THE ATOMIC MASS FORMULA

One of the authors (Cameron, 1959) discussed the abundance distribution of the products formed in the Mike fusion explosion, and reached the conclusion that all mass formulas of conventional form must be poor in the neutron-rich region away from the valley of beta stability. The bases for this conclusion were: (1) the abundance yields produced in the Mike experiment indicated that the neutron capture cross sections of the heavy uranium isotopes could not vary rapidly with mass number, and (2) if the conventional mass formulas are correct in predicting that the neutron binding energy will drop by a factor of 2 or more as 17 neutrons are added to U^{238}, then the theory of neutron capture cross sections indicates that the capture cross sections should drop by a much larger factor in this range.

The argument leading to the conclusion that the cross sections cannot fall off rapidly with increasing mass number (A) is straightforward. Assuming that the abundance of a given isotope is not substantially reduced by neutron capture forming the next isotope, the ratio of the number densities of two successive isotopes is given by

$$\frac{N_i}{N_{i-1}} = \frac{\sigma_{i-1}\Phi}{i}$$

where i denotes the ith capture product in the capture chain, σ_{i-1} is the capture cross section for the $i - 1$ isotope, and Φ is the integrated neutron flux. If the ratios of the yields of successive isotopes remain constant with increasing mass number, and if the flux is uniform, then the cross sections cannot fall off rapidly as predicted by conventional mass formulas. The constancy of the ratios of the yields of successive isotopes for the Mike device is apparent in Figure 1. Similar results have been obtained in recent experiments (Bell, these proceedings).

In an attempt to understand the general features of these yield curves, Cameron and Elkin (1965) were led to a re-evaluation of the atomic mass formula. The nuclear mass excess in MeV on the C^{12} scale of masses is

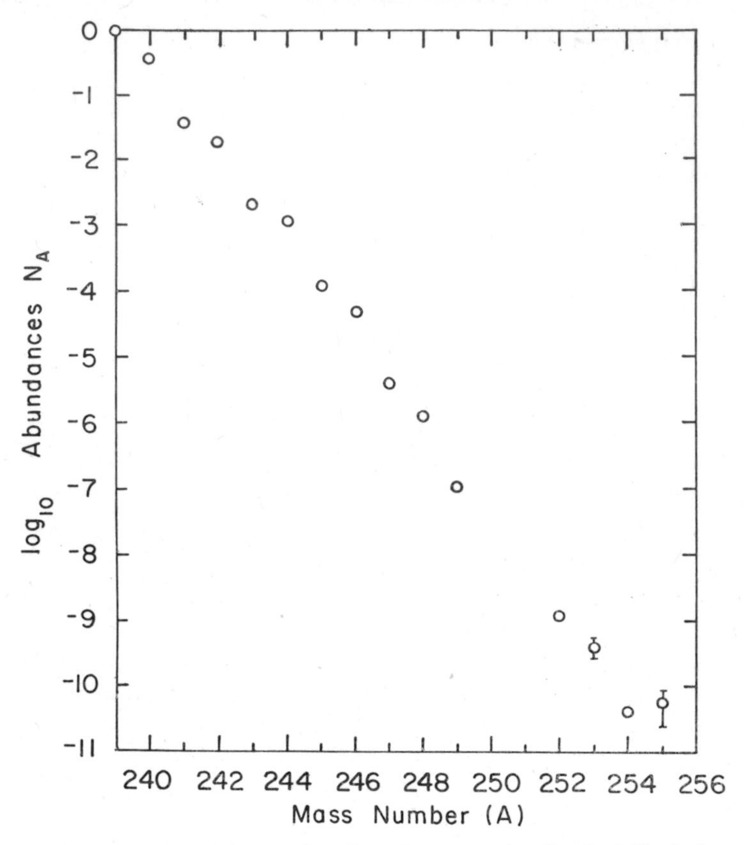

Fig. 1. The yields of nuclides as a function of mass number for the Mike fusion explosion.

written in the form

$$M - A = 8.07134A - 0.78261Z + E_{vs} + E_c + E_{ex} + S(Z, N) + P(Z, N)$$

where E_{vs} includes both volume and surface corrections, E_c is the Coulomb energy correction, E_{ex} is the Coulomb exchange term, and $S(Z, N)$ and $P(Z, N)$ are the total shell and pairing corrections for protons and neutrons.

In a previous consideration of the mass formula (Cameron, 1957) the volume and surface energies, each containing a symmetry term, were written in the form:

$$E_v = \alpha \left[1 - \frac{\beta}{\alpha} \frac{(A - 2Z)^2}{A^2} \right] A$$

$$E_s = \gamma \left[1 - \frac{\varphi}{\gamma} \frac{(A - 2Z)^2}{A^2} \right] \left[\frac{1 - 0.62025}{A^{2/3}} \right]^2 A^{2/3}$$

These terms contain four adjustable parameters, α, β, γ, and φ. The additional factor in the surface energy resulted from a definition of the nuclear radius as that point in a trapezoidal density model of the nucleus at which the density had fallen to half of its central value. The coefficients β and φ are obtained by fitting the position of the valley of beta stability by a least-squares procedure, and the coefficients α and γ are then obtained by making a least squares fit to nuclear masses.

In their study of the mass formula, Cameron and Elkin (1965) attempted to obtain similar coefficients by separate fits to the middle range of mass numbers and to the high range of mass numbers. Similar values were obtained for α, β, and γ, but the coefficient of the surface symmetry term was found to vary enormously. It was concluded that a physically meaningful coefficient for the surface symmetry energy could not be obtained. The combined volume and surface energies were therefore written in terms of three adjustable parameters as

$$E_{vs} = E_v\left[1 + \frac{\gamma}{\alpha A^{1/3}}\left(1 - \frac{0{\cdot}631}{A^{2/3}}\right)^2\right]$$

where, for the conventional form, E_v is given as in the previous case.

The values of the coefficients α and β were determined by Cameron (1957) to be

$$\alpha = -17{\cdot}0354 \text{ MeV.}$$

$$\beta = -31{\cdot}4506 \text{ MeV.}$$

The usual expression for the volume energy would then predict that the binding energy per nucleon will change from -17.0354 MeV for $A = 2Z$ to 14.4512 MeV for a pure neutron gas. Theory predicts, however, that a pure neutron gas should be only slightly unbound (Salpeter, 1960).

Cameron and Elkin (1965) have suggested a form for the volume energy term which will reproduce the conventional results in the region $A \sim 2Z$ while simultaneously leaving pure neutron matter slightly bound, viz:

$$E_v = \alpha A \exp\left[-\frac{\beta}{\alpha}\frac{(A - 2Z)^2}{A^2}\right]$$

This will have the effect of increasing the neutron binding energies for the neutron rich isotopes far from the valley of beta stability. The fact that pure neutron matter is still slightly bound suggests that this correction has gone too far, and that a more realistic extrapolation toward the neutron rich region lies between these two extremes.

There are thus two forms of the mass formula which must be considered, the conventional form and the exponential form, differing only in the definition of the volume energy term E_v. The parameters α, β, and γ are determined by fitting the valley of beta-stability and by fitting masses. Once these have been determined, the shell and paring corrections are found from a study of the differences between the reference mass formula and the known masses. These were determined in such a manner that both S and P are regarded as independent functions of N and Z:

$$S = S(N) + S(Z)$$

$$P = P(N) + P(Z)$$

The fit to the valley of beta stability of the exponential mass formula is shown in Figure 2. This shows the charge difference between the calculated valley of beta stability and the observed valley of beta stability as drawn on

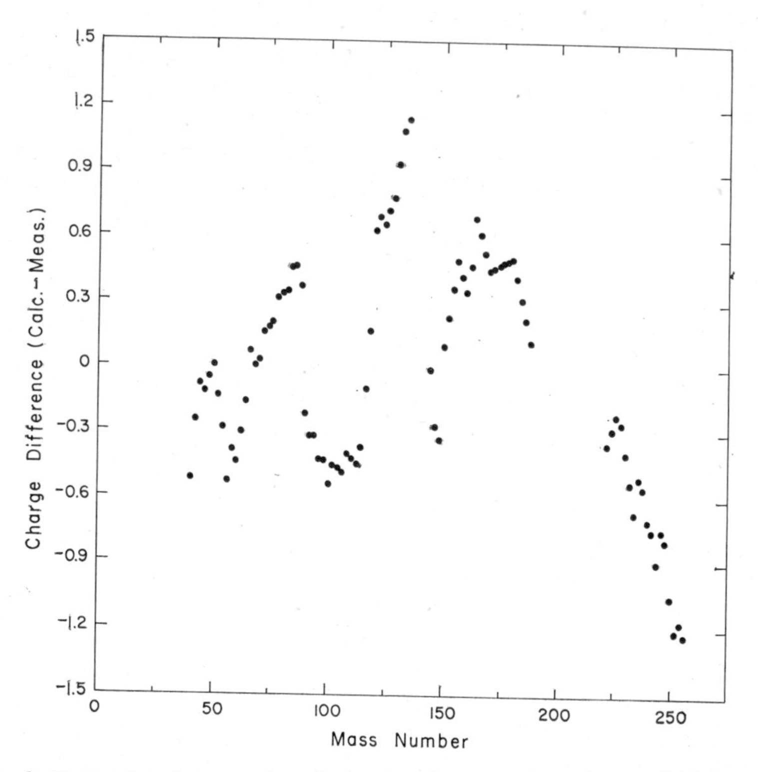

Fig. 2. Comparison between the calculated and measured number at which the valley of beta stability exists for the exponential form of the mass formula.

the General Electric Chart of the Nuclides (Stehn and Clancy, 1956). Although there is generally some scatter about the zero line, there seems to be a systematic departure in the heavy mass region. Figure 3 shows a similar fit for the conventional mass formula. Here again we observe a strong departure from the zero line, but in the opposite sense. Hence, it is evident that an intermediate form of representation of the symmetry energy would probably improve not only the calculation of the neutron binding energies, but also the fit to the valley of beta stability.

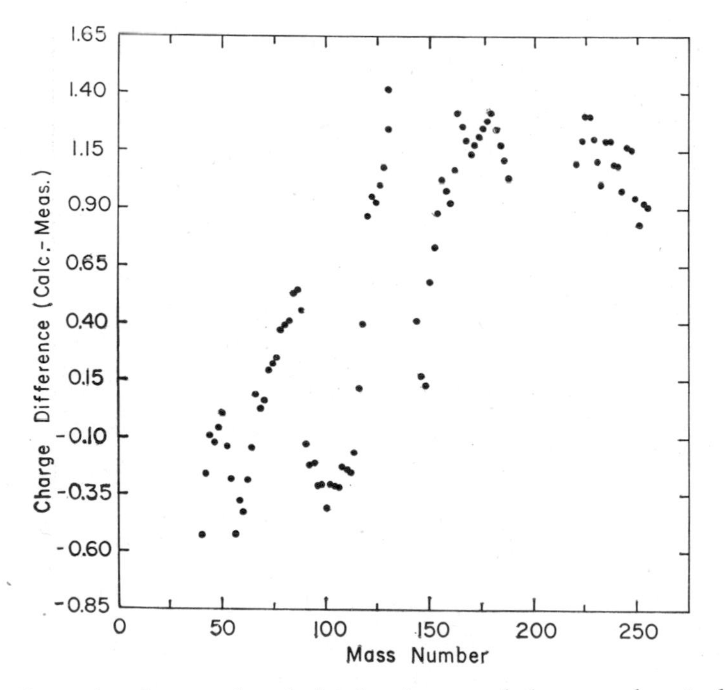

Fig. 3. Comparison between the calculated and measured charge number at which the valley of beta stability exists for the conventional form of the mass formula.

A detailed discussion of the determination of the shell and pairing energies is given by Cameron and Elkin (1965). This procedure is complicated by the fact that one is using a large number of adjustable parameters to obtain a mass formula. The only justification for doing so is that there are about four times as many pieces of experimental data as adjustable parameters, so the system is over-determined to some extent. The first backward differences for the neutron shell corrections, $\delta S(N)$, are plotted in Figure 4. The corresponding curve for protons is shown in Figure 5. These curves effectively represent the slopes of these shell corrections. These slopes are expected to be large and positive immediately following closed shells and to fall gradually

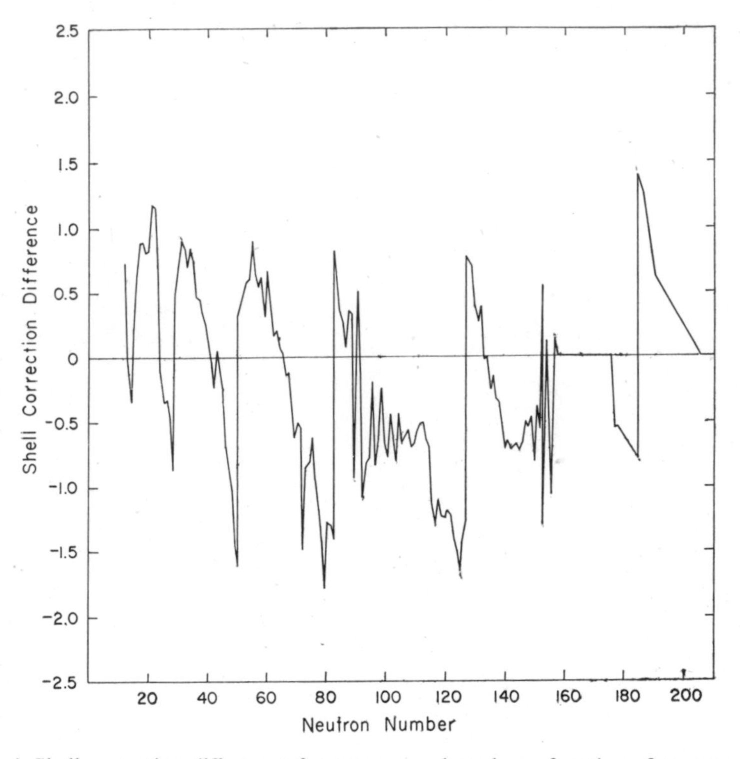

Fig. 4. Shell-correction differences for neutrons, plotted as a function of neutron number, for the exponential form of the mass formula.

to large negative values immediately preceding the next closed shells, as long as the nucleus is undeformed. For a deformed nucleus, $\delta S(N)$ is expected to be small. We observe from Figures 4 and 5 that these expectations are fulfilled and that the shell correction differences are remarkably smooth. This suggests that the procedure used to determine the shell and pairing corrections is a reasonable one.

A comparison of the differences between the mass excesses predicted by the two formulas and the experimental mass excess is shown in Figure 6, together with the errors in the experimental mass excesses. The comparisons are made for four ranges of mass number. Since there are nearly four times

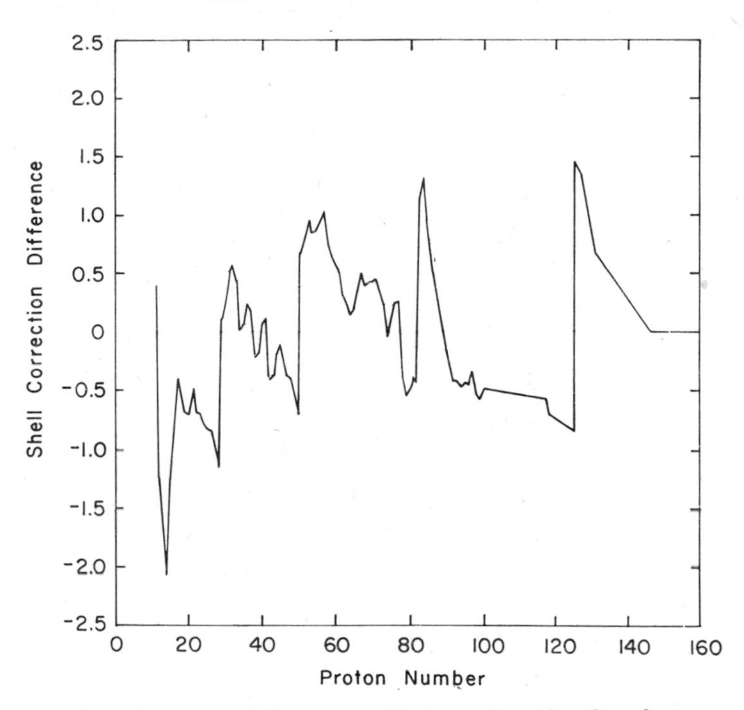

Fig. 5. Shell-correction differences for protons, plotted as a function of proton number for the exponential form of the mass formula.

as many measured mass excesses as adjustable constants in the shell corrections, these comparisons have physical meaning. It may be seen that the differences are significantly larger than the experimental errors, but that there is no significant preference for either the exponential or the conventional formula.

THE NUCLEAR LEVEL DENSITY

Perhaps the most critical factor in determinations of thermonuclear reaction rates is the nuclear level density. Gilbert and Cameron (1965) have recently examined this subject in some detail. From a study of the known levels at low excitation energies and of the behavior of the level density in the vicinity of the neutron binding energy they have determined parameters defining a fit to the observed level densities.

The original formulation of the statistical theory of nuclear level densities is due to Bethe (1937). In this theory the nucleus is considered to be a fermi

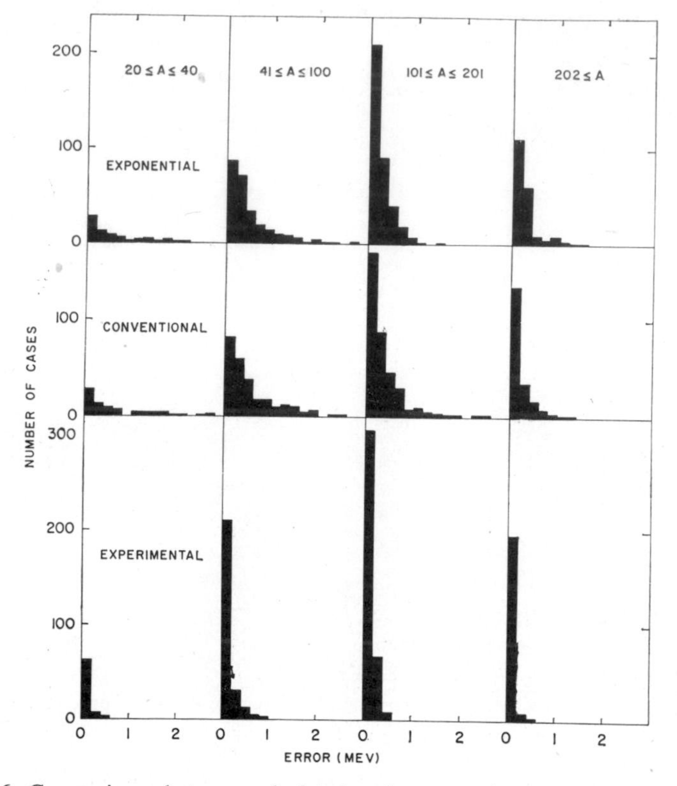

Fig. 6. Comparisons between calculated and measured mass excesses. The absolute values of the differences for the exponential case are plotted on the top row for four ranges of mass number; those for the conventional case are in the middle row. The bottom row shows the distribution of measured errors.

gas composed of two types of fermions. By treating the number of protons, the number of neutrons, the total energy, and the total magnetic quantum number as constants of the motion, one can define the grand canonical partition function for the system. Proceeding in this manner Gilbert and Cameron (1965) arrive at the following expression for the density of states of all possible magnetic quantum numbers, M:

$$\omega(U) = \frac{\pi^{1/2}e^2\, 2^{(aU)^{1/2}}}{12a^{1/4}\, U^{5/4}}$$

In this expression

$$a = \frac{\pi^2}{6} g$$

where g^{-1} is the single particle level spacing:

$$U = E - U_0$$

is the effective excitation energy; and U_0 is the Fermi energy.

$\omega(U)$ represents the total density of states, including states degenerate in the magnetic quantum number M. We require, rather, an expression for the density of levels of given angular momentum J, parity π and energy U. Following Bethe (1937) the density of levels of specified J can be written in the form:

$$\varrho(U, J) \simeq \frac{\omega(U)}{(2\pi\sigma^2)^{1/2}} \frac{(2J+1)}{2\sigma^2} \varepsilon^{-(J+1/2)^2/2\sigma^2}$$

In the absence of experimental evidence favoring even or odd parity states we will assume there is equal probability for either parity. The density of levels of spin J, parity π at an excitation energy U is thus given by:

$$\varrho(U, J, \pi) = \frac{1}{(12)\, 2^{1/2}} \frac{e^{2(aU)^{1/2}}}{a^{1/4} U^{5/4} \sigma} \left\{ \frac{(2J+1)}{2\sigma^2} e^{-(J+1/2)^2/2\sigma^2} \right\} \left\{ \frac{1}{2} \right\}$$

In this expression σ is the spin dependence parameter given by

$$\sigma^2 = g\langle m^2 \rangle\, t$$

where t is the nuclear temperature

$$t = \left(\frac{U}{a} \right)^{1/2}$$

and $\langle m^2 \rangle$ is the mean-square single particle magnetic quantum number. Jensen and Luttinger (1952) have demonstrated that the succession of shell-model states implies

$$\langle m^2 \rangle \sim 0.146 A^{2/3}$$

with some fluctuations attributable to shell effects. Thus we find:

$$\sigma^2 = 0.0888\, (aU)^{1/2} A^{2/3}$$

In this formula there are in effect two free parameters, a and U_0. It was initially assumed that U_0, the energy of the fully degenerate state, represented the ground state of the nucleus. 'a' could then be determined from experimentally known neutron resonance spacings. However, it was found that there would be systematic differences in the values of 'a' for neighboring even-even, odd A, and odd-odd nuclei. Thus, a pairing correction is necessary. Such a correction is in fact supplied by the mass formula: if U_0

is taken to be the 'pairing energy' it is found that these odd-even effects can be taken into account.

The Bethe formula is not expected to valid at low excitation energies. Ericson (1959) has shown that the expression

$$\varrho(E) = \frac{e^{(E - E_0)/T}}{T}$$

provides a good fit to empirically determined nuclear levels over the first few MeV of excitation. The density of levels of specified angular momentum and parity is then given by

$$\varrho(E, J, \pi) = \varrho(E)\left\{\frac{(2J + 1)}{2\sigma^2}\, e^{-(J + 1/2)^2/2\sigma^2}\right\}\left\{\frac{1}{2}\right\}$$

Typical examples of the agreement of the level density behavior at low excitations with the Ericson formula are shown in Figure 7. Here, the number of levels up to some excitation energy, E, is plotted as a function of energy. The parameters E_0 and T can be determined from the intercepts and slopes of the straight lines drawn in this figure.

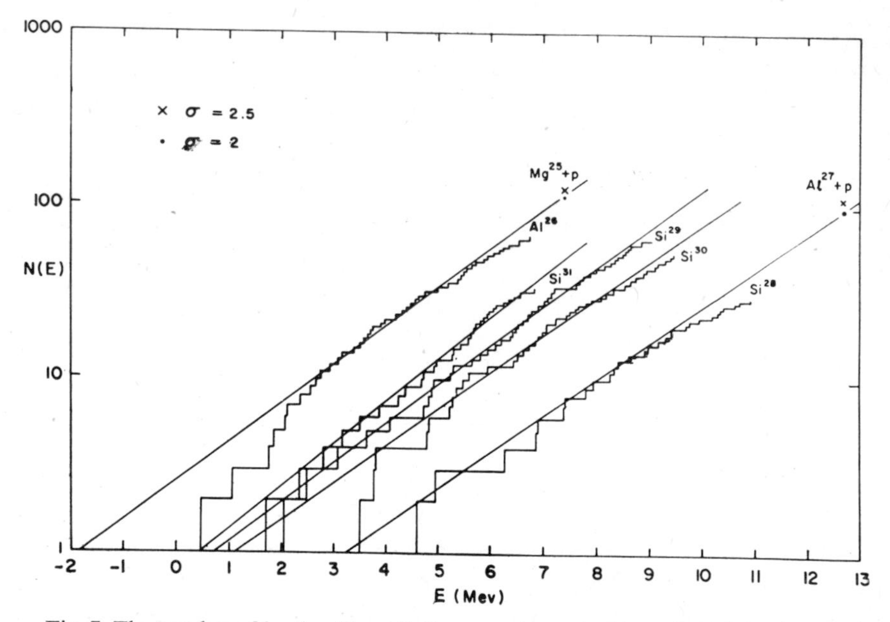

Fig. 7. The number of levels with excitation energies up to E are plotted for the indicated nuclei. The level densities inferred from proton resonance densities in the vicinity of the proton binding energy are shown for two values of the spin parameter, σ,

It seems reasonable to ask whether a dependence like $\exp(a(E - E_0))^{1/2}$ is a better representation of the levels near ground. This can be tested rather precisely. Figures 8 and 9 show the results, for Cl^{36}, of plotting $\log^2 N(E)$ and $\log N(E)$, respectively, as a function of energy. For both cases, straight line firs to the data can be obtained, On the basis of the low-lying levels alone, it would seem that an $\exp(a(E - E_0))^{1/2}$ dependence provides a better fit. However, this form is far off in the vicinity of the neutron resonances. This conclusion is not dependent upon the choice of the spin parameter, σ. It also holds for other nuclei for which both lower levels and neutron resonances are known.

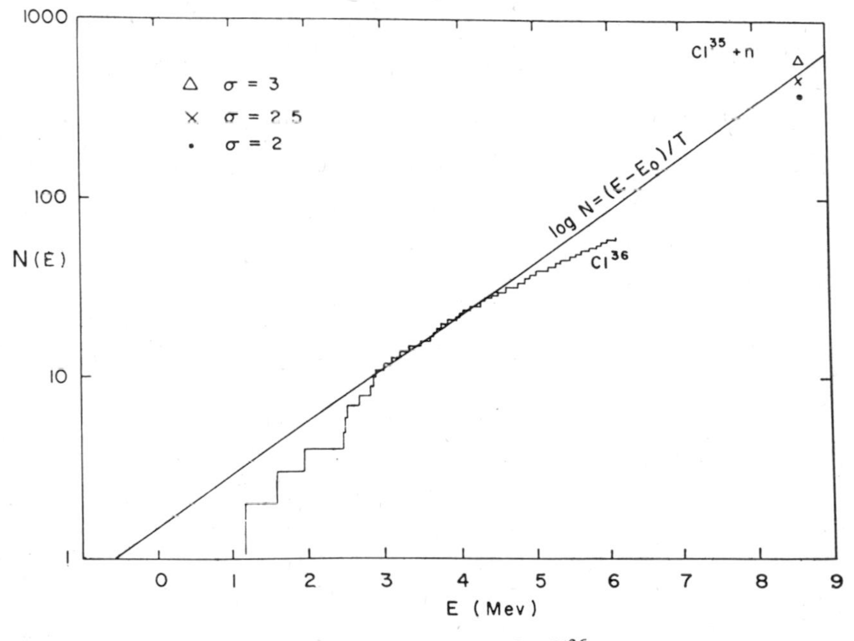

Fig. 8. Log $N(E)$ vs. E for Cl^{36}.

Gilbert and Cameron (1965) have shown that a reasonable and self-consistent description of the level density at any excitation can be obtained by using the Bethe formula at high energies, the Ericson formula at low energies, and fitting the two forms tangetially. The parameter a can be determined by studies of the level densities near the proton and neutron binding energies. While the treatment of the nucleus as a Fermi gas predicts $a \propto A$, this does not allow us to take into account the obvious shell effects in

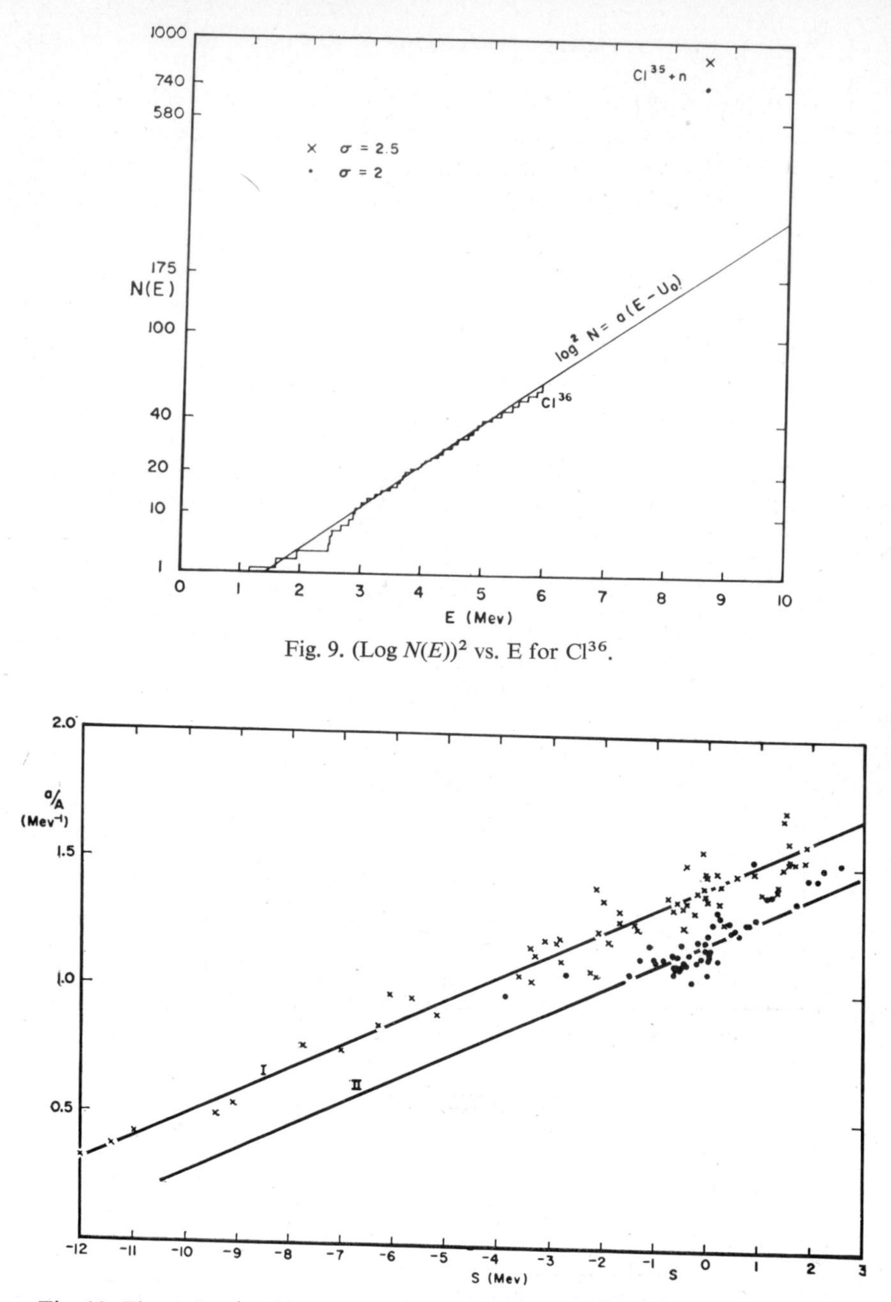

Fig. 9. $(\text{Log } N(E))^2$ vs. E for Cl^{36}.

Fig. 10. The ratio a/A plotted as a function of the total shell correction S for nuclei heavier than zinc. Crosses correspond to undeformed nuclei, dots to deformed nuclei. The lines drawn through these data determine the correlations for deformed and undeformed nuclei.

neutron resonance spacings. Gilbert and Cameron (1965) determined the following correlations for the parameter a, for undeformed nuclei

$$\frac{a}{A} = 0.00917S + 0.142\ (\text{MeV}^{-1})$$

and for deformed nuclei

$$\frac{a}{A} = 0.00917S + 0.120\ (\text{MeV}^{-1})$$

Here S is the total shell correction, the sum of the shell corrections for neutrons and protons. These correlations were inferred from the data shown in Figures 10, 11, and 12. The shell corrections are those predicted by the exponential form of the mass formula.

These correlations suggest that one can calculate the level densities at high excitations knowing the pairing and shell corrections. The ratios of the calculated densities, from the exponential mass formula parameters, to the observed level densities at the neutron binding energies are plotted in

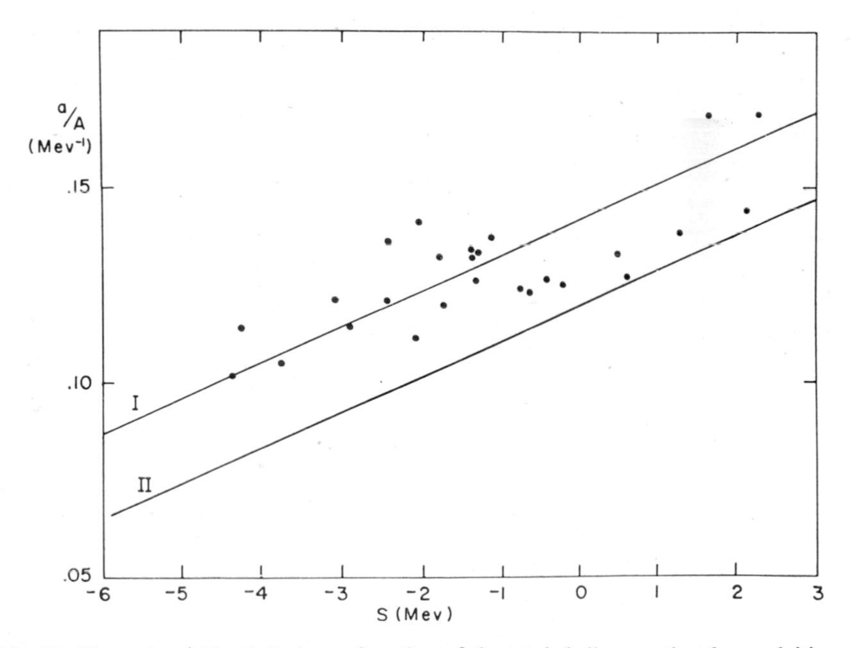

Fig. 11. The ratio a/A is plotted as a function of the total shell correction for nuclei in range calcium to zinc. Generally, the nuclei seem to prefer line I for undeformed nuclei.

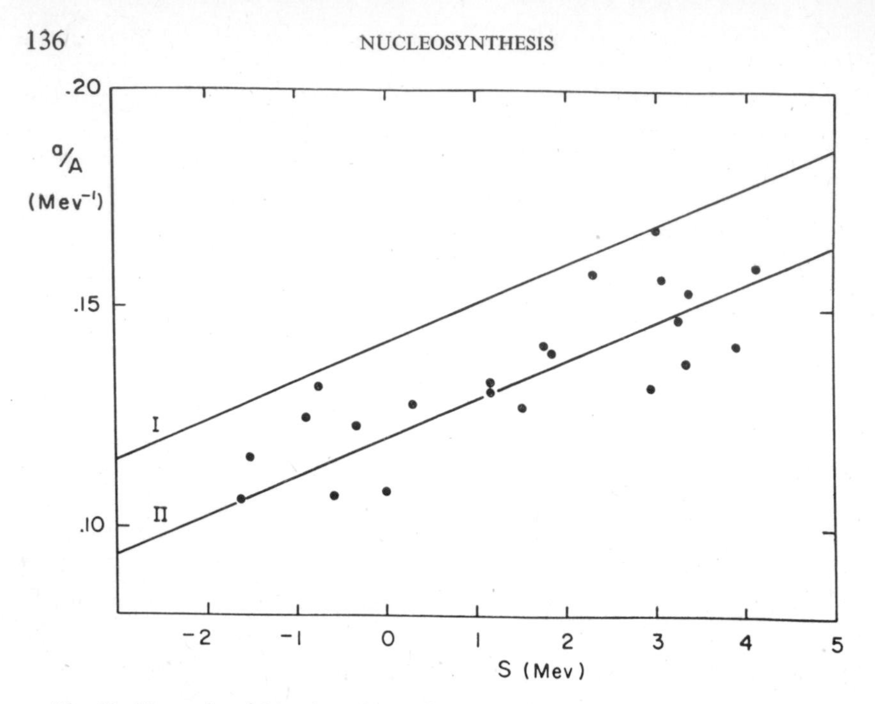

Fig. 12. The ratio a/A is plotted as a function of the total shell correction for nuclei below calcium. While these nuclei tend to scatter about the line for deformed nuclei, this may represent a failure of the mass formula to give good shell corrections in this mass range.

Figure 13 for nuclei above zinc. This is a scatter diagram, showing the errors involved in making such predictions. The probable error for the calculated values is a factor of 1.75.

The parameter T is determined by fitting the levels near ground. The behavior of T as a function of mass number, A, is shown in Figure 14. Some shell effects are evident in this case, as are known to exist.

The behaviour of the intercept, E_0, as a function of mass number is shown in Figure 15. A correlation of the intercepts with the pairing energy is evident in this figure Even-even nuclei tend to lie close to the positive pairing energy correlation line and odd-odd nuclei near the bottom line, while the points for odd-A nuclei are scattered about zero.

The fitting energies, U_x, determined in this investigation are plotted as a function of mass number in Figure 16. This dependence is expressed approximately by

$$U_x = 2.5 + \frac{150}{A}\ \text{MeV}$$

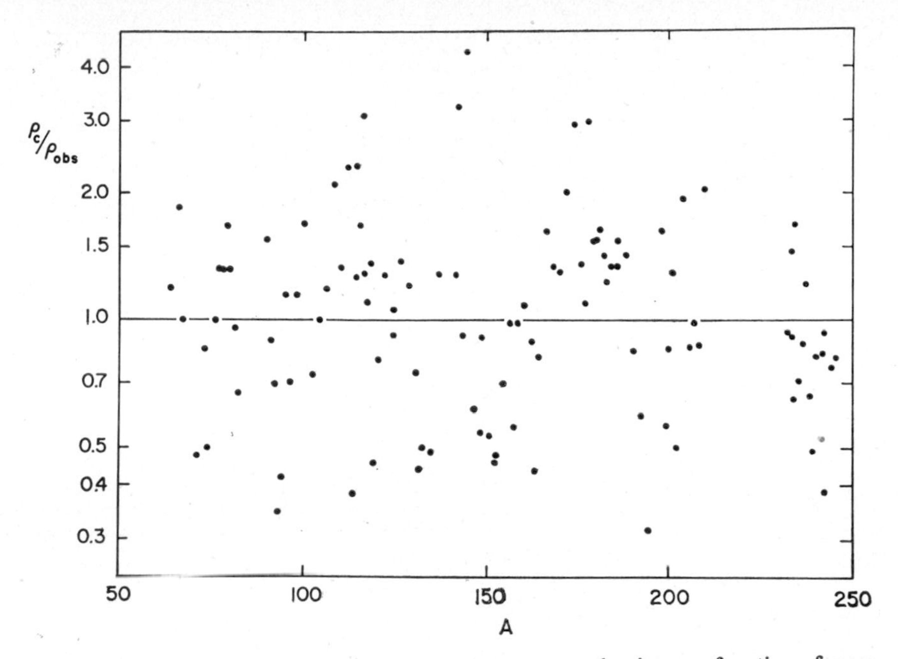

Fig. 13. Ratio of calculated to observed neutron resonance density as a function of mass number.

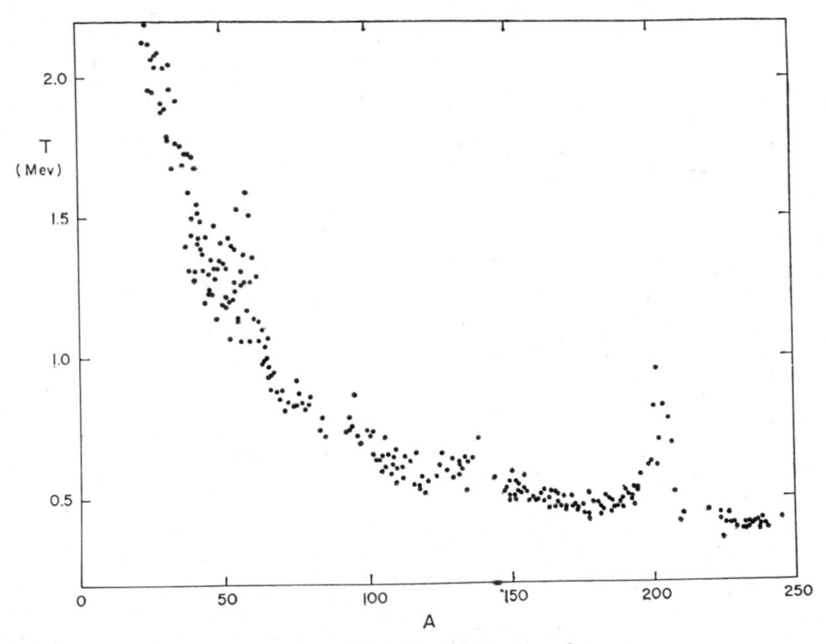

Fig. 14. The nuclear temperature T as a function of mass number.

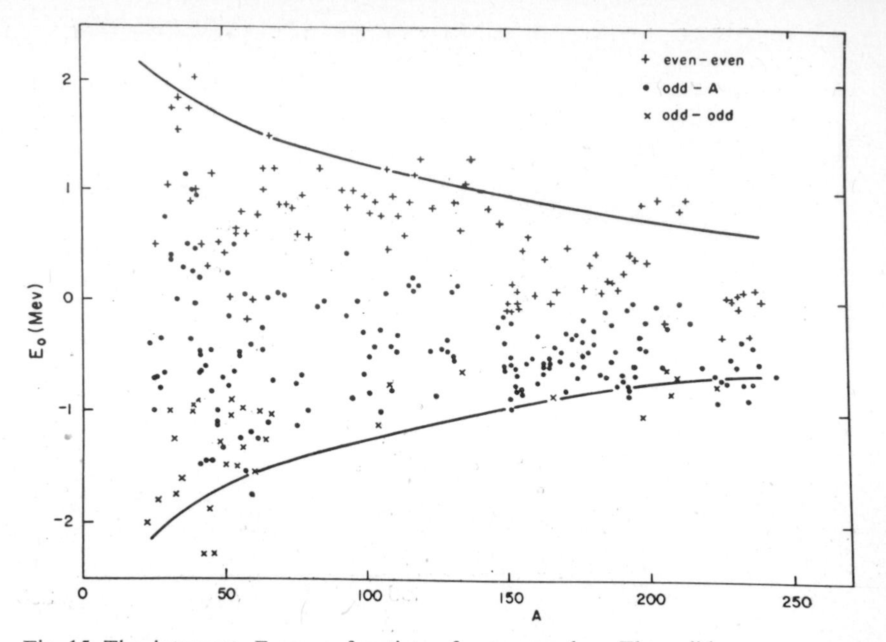

Fig. 15. The intercept E_0 as a function of mass number. The solid curves represent $\pm \langle P \rangle$, the average pairing energy.

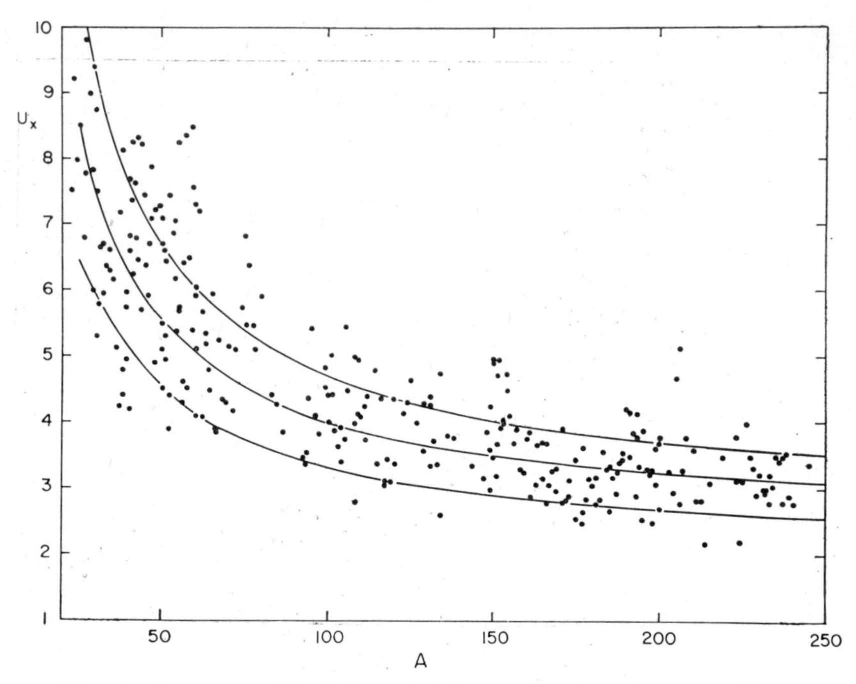

Fig. 16. The energy of the fitting point U_x as a function of mass number.

In fact, the fitting is not extremely sensitive to U_x. This is illustrated in Figure 17, for the Sm^{148} nucleus. It is clear that the fitting energy U_x can be moved up and down somewhat without significantly changing the manner in which the fitting occurs.

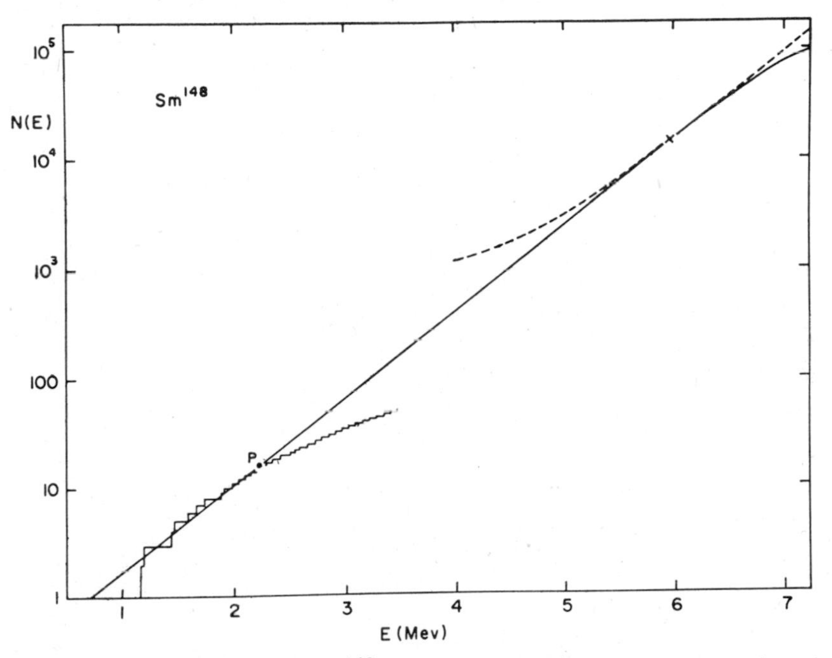

Fig. 17. Number of level of Sm^{148} up to energy E. The cross represents the point of tangency. The dotted lines are extrapolations of the low-energy formula above this point and of the high-energy point below it.

DETERMINATIONS OF THE WIDTHS

In order to determine thermonuclear reaction rates, it was necessary to calculate both radiation widths and particle widths from these statistical nuclear properties.

Following Blatt and Weisskopf (1952) we can write the total width for electric-dipole radiation from a level of energy E_a, spin J, and parity α in the form (generally, electric dipole transitions are dominant)

$$\Gamma_\gamma(E_a, J, \pi) = 0.296 \frac{A^{2/3}}{D_0} \sum_{J', \pi'} \frac{1}{\varrho(E_a, J, \pi)} \int_0^{E_a} E^2 \varrho(E_a - E, J', \pi') \, dE \tag{5.1}$$

where D_0 is effectively a normalization factor, although it may be interpreted in some sense as the single particle level spacing. Assuming an equality of the parity distributions and neglecting the exponential term in the angular momentum dependence of the level density, the summation is performed over all allowed values of J', π' of the daughter levels consitent with electric dipole selection rules ($\Delta J = \pm 1, 0$; not $0 \to 0$; parity change). This introduces a factor

$$\frac{[2(J+1)+1]+[2(J-1)+1]+[2J+1]}{(2J+1)} = \frac{6J+3}{2J+1} = 3$$

into the expression for the width; hence:

$$\Gamma_\gamma\,(E_a, J, \pi) = \frac{0.89A^{2/3}}{D_0}\,\frac{1}{\varrho(E_a)}\int_0^{E_a} E^3\varrho(E_a - E)\,dE$$

The widths calculated from this expression, employing the level density parameters from Gilbert and Cameron (1965), have been compared with experimental values for a large number of nuclei with $A > 40$, where it was required that at least the value of Γ_γ for one level in the nucleus had been definitely established. These experimental values were taken from the Nuclear Data Sheets and from Hughes et al. (1960). The value of D_0 inferred from these calculations was $D_0 \sim 230$ MeV. Calculations performed for $A \lesssim 40$ led to a value $D_0 \simeq 400$ MeV.

The ratio Γ_γ (calculated)/Γ_γ (experimental) from these calculations is plotted in Figure 18. The experimental errors associated with the values of Γ_γ are generally $\lesssim 30\%$. The dashed line in this figure is a 'guide to the eye' for this ratio. The solid line corresponds to the optical model, deformed nucleus s-wave neutron strength function due to Chase *et al.* (1958). The general behavior of these two curves is quite similar for $A \gtrsim 80$.

From nuclear reaction theory, the particle widths can be written in the form

$$\Gamma_l = 2P_l\gamma_l^2$$

where l specifies the value of the orbital angular momentum and P_l is the nuclear penetrability. The factor γ_l^2 is the reduced width for the particle channel and is dependent upon the behavior of the wave function within the nucleus.

Generally a detailed knowledge of the form of the reduced width is not available. We have assumed in our calculations that the strength functions

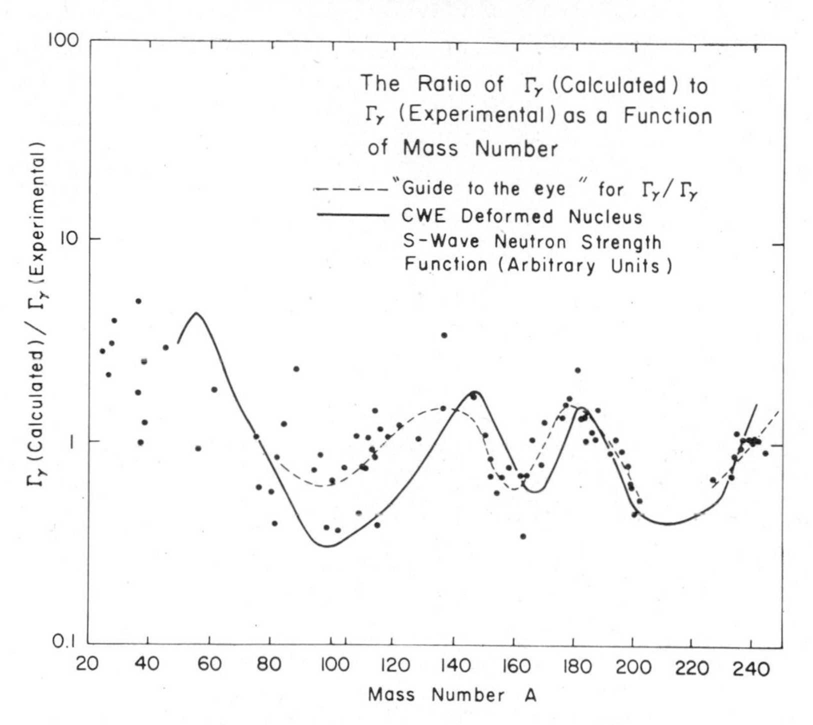

Fig. 18. The ratio of Γ_γ (calculated) to Γ_γ (experimental) is plotted as a function of mass number.

for the various partial waves for protons, neutrons, and alpha-particles can be approximated by the strength function predicted by a 'black nucleus' model. On this model γ_l^2 can be written in the form

$$\gamma_l^2 = \frac{2 \times 10^{-14}}{R\varrho(E, J, \pi)}$$

where R is the nuclear radius and $\varrho(E, J, \pi)$ is the level density. The nuclear radius is determined from an expression of the form

$$R = r_0(A_P{}^{1/3} + A_T{}^{1/3})$$

where A_P and A_T are the mass numbers of the projectile and target nucleus respectively.

REACTION RATE CALCULATIONS

Our consideration of these various nuclear properties has been undertaken in an attempt to calculate accurate values for thermonuclear reaction rates

(Truran *et al.* 1966). These rates can be computed from equation (1), for appropriate values of the level densities, the particle widths, and the radiation widths.

Experimental determinations of the individual resonance parameters are available for a number of alpha-particle and proton reactions proceeding on light nuclei. This allows a determination of the $\langle \sigma v \rangle$ as a sum over the contributions from individual resonances. The ratios of our calculated values of $\langle \sigma v \rangle$ to $\langle \sigma v \rangle$ determined from experimental parameters are plotted as a function of temperature in Figure 19. Discrepancies of a factor of two or three of our rates from the experimental values can readily be

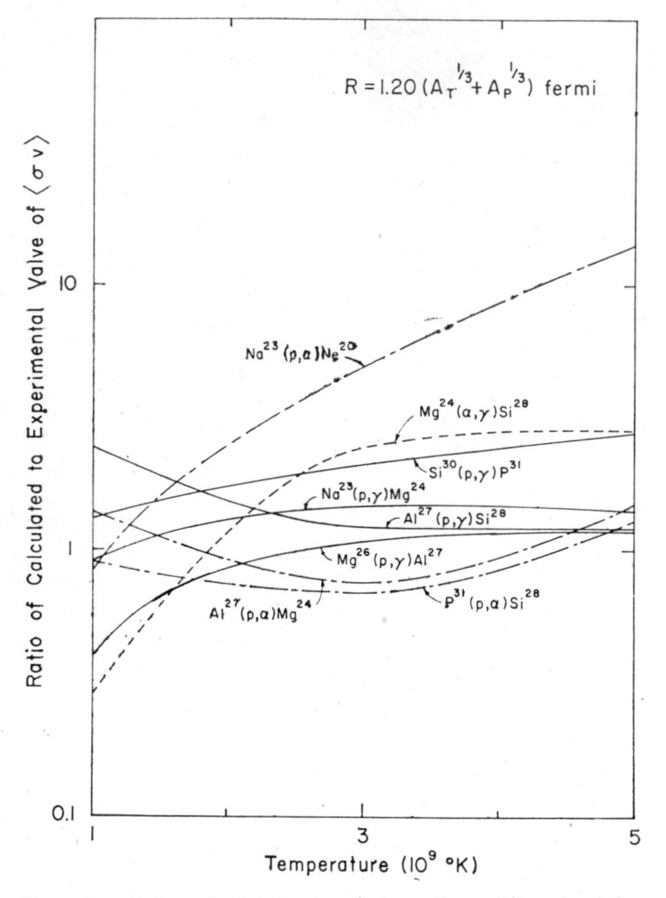

Fig. 19. The ratio of the calculated rates $\langle \sigma v \rangle$ to those determined from experimental resonance parameters are plotted as function of temperature for a value of the nuclear radius parameter $r_0 = 1.20$.

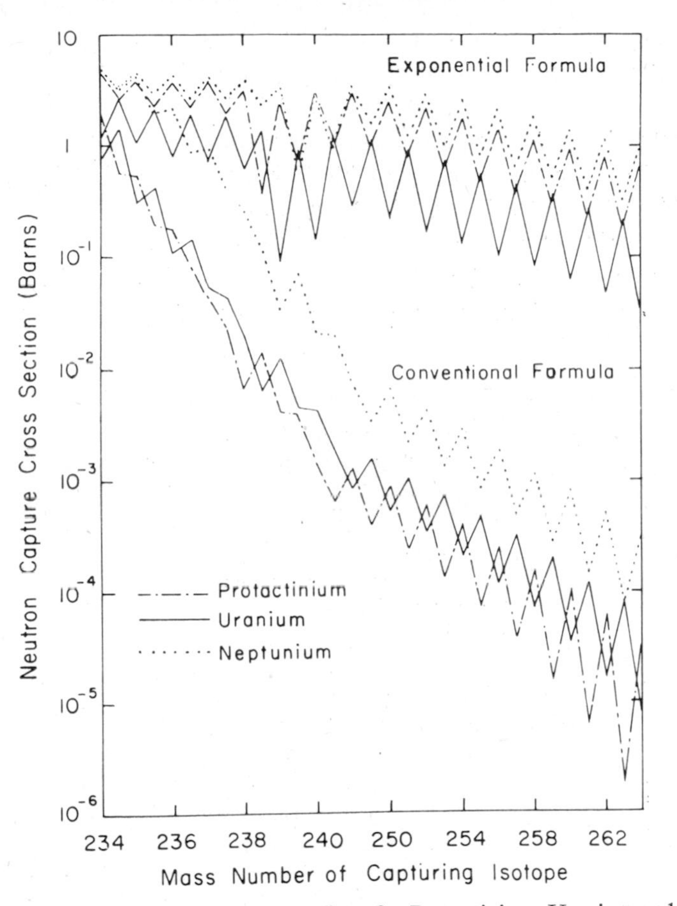

Fig. 20. The neutron capture cross sections for Protactinium, Uranium and Neptunium isotopes, calculated from both the exponential and conventional mass formula parameters, are plotted as a function of mass number. The energy at which these cross sections have been evaluated is 20 keV.

accounted for through the uncertainties in our values of the particle and radiation widths and in the level densities. The large deviation of $Na^{23}(p, \alpha)$ is due to the fact that there are large contributions to the integral from resonances at energies higher than the maximum energy of the experimental resonances. The other reactions give satisfactory results.

With regard to our earlier discussions of the yield curve for the Mike explosion, we have calculated the neutron capture cross-sections for a number of heavy isotopes of protactinium, uranium, and neptunium. These calculations have been performed for the predictions of both the conventional and exponential mass formulas for the shell corrections, and the neutron binding energies. The results of these calculations, for an energy of 20 keV, are shown in Figure 20. In these calculations, we have assumed that the total width is the sum of the neutron width and the radiation width, decay into other channels being negligible. We have also assumed a spin 5^+ for odd-odd nuclei, and $5/2^+$ for odd A nuclei.

The cross sections for these isotopes exhibit the usual odd-even effect. Generally we observe that while the cross sections predicted by the conventional mass formula fall off by three or four orders of magnitude between

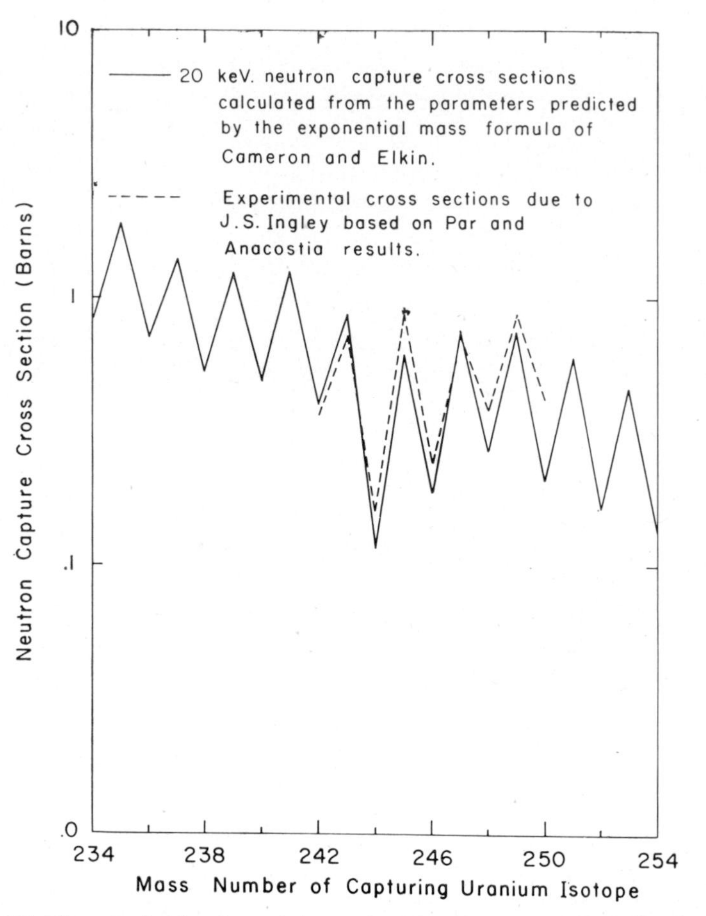

Fig. 21. The calculated values of the neutron capture cross sections for the Uranium isotopes are compared to experimental values inferred from the yields of the Par and Anacostia devices (Ingley, 1965).

mass numbers A = 234 and 264, the capture cross sections determined for the exponential form decrease by roughly a factor of ten. The observed slow decrease in the capture cross sections determined from the exponential mass formula parameters is consistent with the behavior inferred from the yield curves.

The dip in our capture cross sections corresponding to neutron number 152 must be interpreted with caution. There is a gap in the Nilsson diagram for deformed nuclei at this point (Mottleson and Nilsson, 1958) which predicts this effect for heavier nuclei on the valley of beta stability. However, as the mass formula parameters were fitted to nuclei along the beta stable valley the existence of this effect in our calculated cross sections for nuclei in the neutron rich region must be interpreted with caution.

Ingley (1965) has recently determined values for the neutron capture cross sections for a number of uranium isotopes from the results of the Par and Anacostia experiments. His results are plotted in Figure 21 together with our calculated values for the neutron capture cross sections, for neutron energy 20 keV. The neutron width fluctuations have been taken into account in these calculations. The agreement is seen to be quite good. Furthermore, the same dip in the capture cross section at U^{244} is evident in these results, suggesting that the Nilsson gap at neutron number 152 is present in the heavy uranium nuclei.

REFERENCES

Bethe, H. (1937). *Rev. Mod. Phys.* **9**, 69.

Cameron, A. G. W. (1957). *Can. J. Phys.* **35**, 1021.

Cameron, A. G. W. (1959). *Can. J. Phys.* **37**, 322.

Cameron, A. G. W. and Elkin, R. M. (1965). *Can. J. Phys.* **43**, 1288.

Chase, D. M. Wilets, L. and Edmonds, A. R. (1958). *Phys. Rev.* **110**, 1080.

Ericson, T. (1959). *Nucl. Phys.* **11**, 481.

Gilbert, A., Chen, F. S. and Cameron, A. G. W. (1965). *Can. J. Phys.* **43**, 1248.

Gilbert, A. and Cameron, A. G. W. (1965). *Can. J. Phys.* **43**, 1446.

Hayashi, C., Hoshi, R. and Sugimoto, D. (1962). *Suppl. Prog. Theor. Phys.* **22**, 1.

Hughes, D. H., Magurno, B. A. and Brussel, M. K. (1960). Brockhaven National Laboratories Report, NBL-325.

Ingley, T. S. (1965). Private communication.

Jensen, J. H. D. and Luttinger, J. M. (1952). *Phys. Rev.* **86**, 907.

Mottleson, B. R. and Nilsson, S. G. (1959). *K. Danske Vidensk. Selsk. mat.-fys. Skr.* **1**, No. 8.

Salpeter, E. E. (1960). *Ann. Phys.* **11**, 393.
Stehn, J. R. and Clancy, E. F. (1956). *General Electric Chart of the Nuclides* (Knolls Atomic Power Laboratory publication).
Truran, J. W., Hansen, C. J., Cameron, A. G. W. and Gilbert, A. (1966). *Can. J. Phys.* **44**, 151.

The Approach to Nuclear Statistical Equilibrium

A. GILBERT, J. W. TRURAN, and A. G. W. CAMERON

At high temperatures and densities, thermonuclear reactions will proceed rapidly and an equilibrium can be established among the various nuclear species present in the medium. It has generally been assumed that the formation of the iron peak elements—Cr, Mn, Fe, Co, and Ni—can be attributed to this equilibrium process. The general features of the iron abundance peak observed in nature can be reasonably well reproduced in this manner, for $T \sim 4 \times 10^9$ °K and $\varrho \sim 10^6$ gm/cc (Burbidge *et al.*, 1957).

We have sought to follow in detail the approach to this equilibrium configuration. Our choice of initial conditions follows from our knowledge of the early stages of stellar evolution. The details of stellar evolution through the hydrogen and helium burning phases are quite well established (Hayashi *et al.*, 1962). Following helium burning, the destruction of C^{12} can proceed at temperatures $T \geqq 7 \times 10^8$ °K by the reaction $C^{12} + C^{12}$ (Cameron, 1959a; Reeves and Salpeter, 1959). At somewhat higher temperatures ($T \sim 10^9$ °K) oxygen burning by $O^{16} + O^{16}$ is also possible (Cameron, 1959b; Reeves and Salpeter, 1959). The details of these reactions are not well determined. The reaction products should be nuclei with masses in the range $20 \leqq A \leqq 32$, particularly the alpha-particle nuclei Mg^{24}, Si^{28}, and S^{32}. Of the nuclei in this region, Si^{28} has the highest separation energies for protons and alpha-particles. As the temperature is increased, Si^{28} will be the last nucleus to be photodisintegrated and will therefore accumulate. It can be assumed that the material consists mainly of Si^{28} after carbon and oxygen burning have taken place, at temperatures $T \gtrsim 2 \times 10^9$ °K.

At temperatures $\gtrsim 3 \times 10^9$ °K, the photodisintegration of silicon will proceed rapidly, releasing protons, neutrons and alpha-particles. The capture of these light particles on nuclei remaining in this region will result in the buildup of nuclei to the vicinity of iron. These nuclear transformations comprise the equilibrium process. The important consideration is whether or not the evolution of a presupernova star proceeds too rapidly to allow the silicon to iron conversion. This is a function of the extent to which neutrino energy losses will accelerate the evolution. From a consideration of the rate of energy loss by the pair annihilation process

$$e^- + e^+ \to \nu + \bar{\nu}$$

Stothers and Chiu (1962) have found the evolutionary time for $T = 3 \times \times 10^9$ °K and $\varrho = 2 \times 10^6$ gm/cc to be only 0.3 years. It is clear that the occurence of an equilibrium process at an appreciably higher temperature is unlikely.

There is, however, another manner in which these iron peak elements might be synthesized. The formation of a shock wave in the stellar core may result from the collapse of the core (Cameron, 1963). Colgate, Grasberger, and White (1961) have followed the dynamical implosion of a presupernova star of ten solar masses (10 $M_{\odot}$) through the formation of the shock wave. The mass ejected by the subsequent passage of the shock outward through the envelope was 1 $M_{\odot}$. The temperature and density of the medium at this point, immediately following the passage of the shock, was $T = 5 \times 10^9$ °K and $\varrho = 1.3 \times 10^7$ gm/cc. The temperature was found to remain above $\sim 5 \times 10^9$ °K for approximately 10^{-2} seconds, falling by an order of magnitude in the first second.

As we are concerned with the general problem of the production of iron peak elements in stellar interiors, we have followed the silicon-to-iron conversion for two temperatures, $T = 3$ and 5×10^9 °K, corresponding, respectively, to the conditions predicted for the medium following carbon and oxygen burning and to those predicted for the medium in the wake of the shock. In order to follow these nuclear transformations, a nuclear reaction network was established providing suitable reaction links connecting neighboring nuclei (Figure 1). All proton, neutron, and alpha-particle reaction links were included, together with the reactions $C^{12} + C^{12}$, $O^{16} + O^{16}$ and $3\alpha \to C^{12}$. The contributions of nuclear beta decays were also considered. A method of solution of the network equations has been discussed in detail by Truran *et al.* (1966).

The early stages of evolution of the silicon region at $T_9 = 5$ are displayed in Figure 2. The photodisintegration of silicon to alpha-particles results in a rapid buildup of the products Mg^{24}, Ne^{20} and O^{16}. The subsequent capture of alpha-particles on silicon in these early stages leads to a rapid rise in the abundance of S^{32}. C^{12} increases slowly in abundance due to the relatively low value of the rate for O^{16} (γ, α) C^{12}. The alpha-particle nuclei are

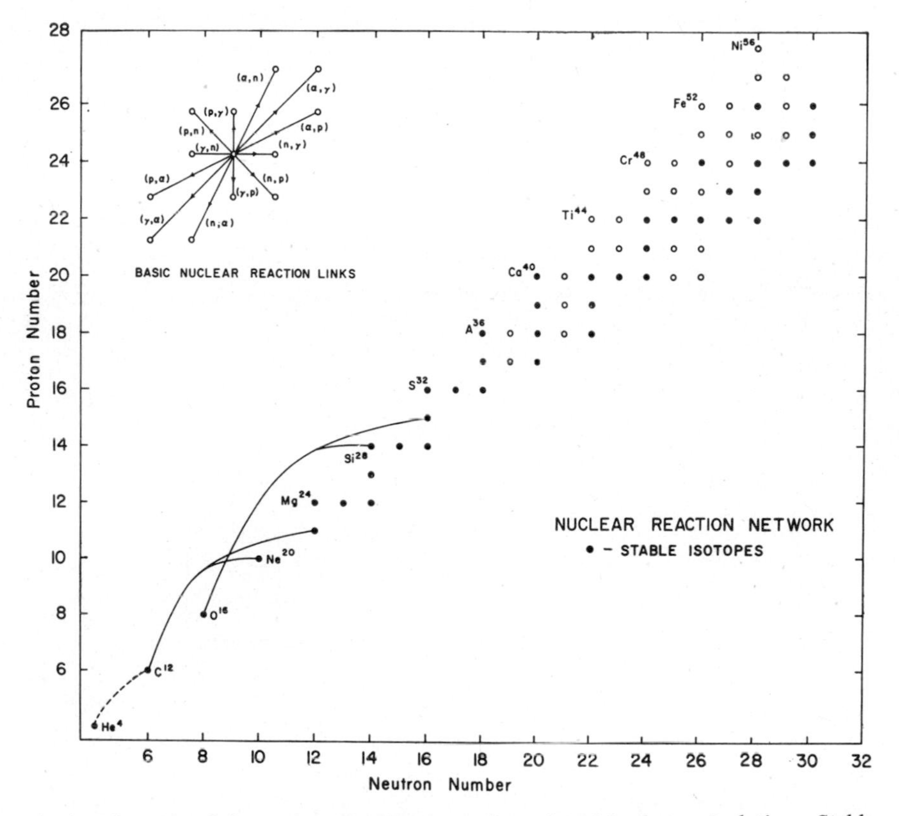

Fig. 1. Schematic of the nuclear reaction network employed in these calculations. Stable isotopes are designated by black dots; unstable isotopes by open circles.

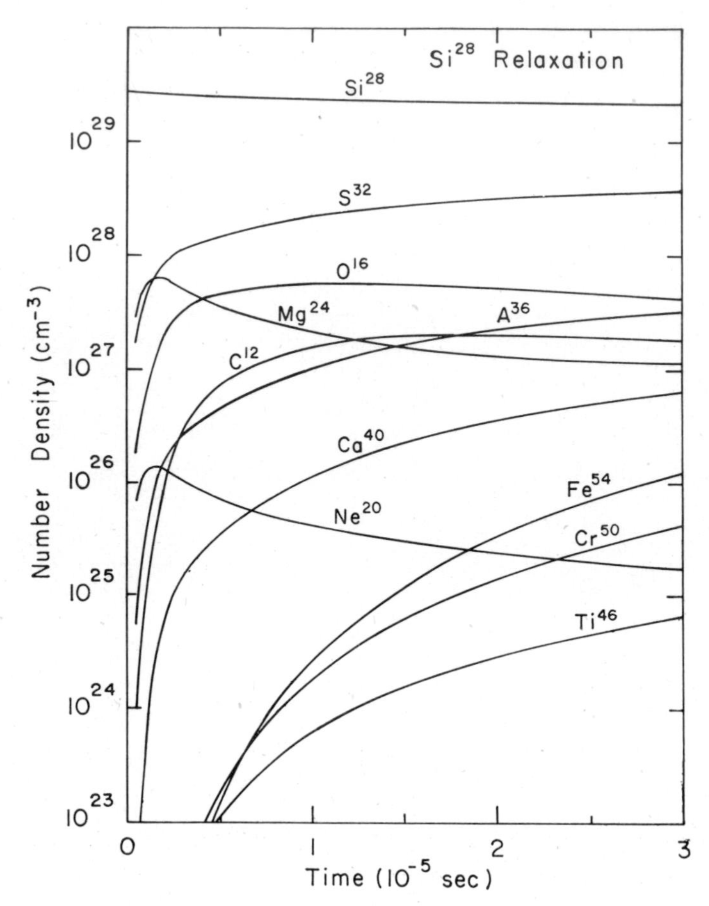

Fig. 2. The early stages of evolution of a region composed initially of pure Si^{28} at a temperature $T_9 = 5$.

generally the most abundant products in their respective mass ranges through Ca^{40}.

Figures 3 and 4 show the late stages of evolution of this silicon region. As the alpha-chain nuclei come into relative equilibrium with silicon, their

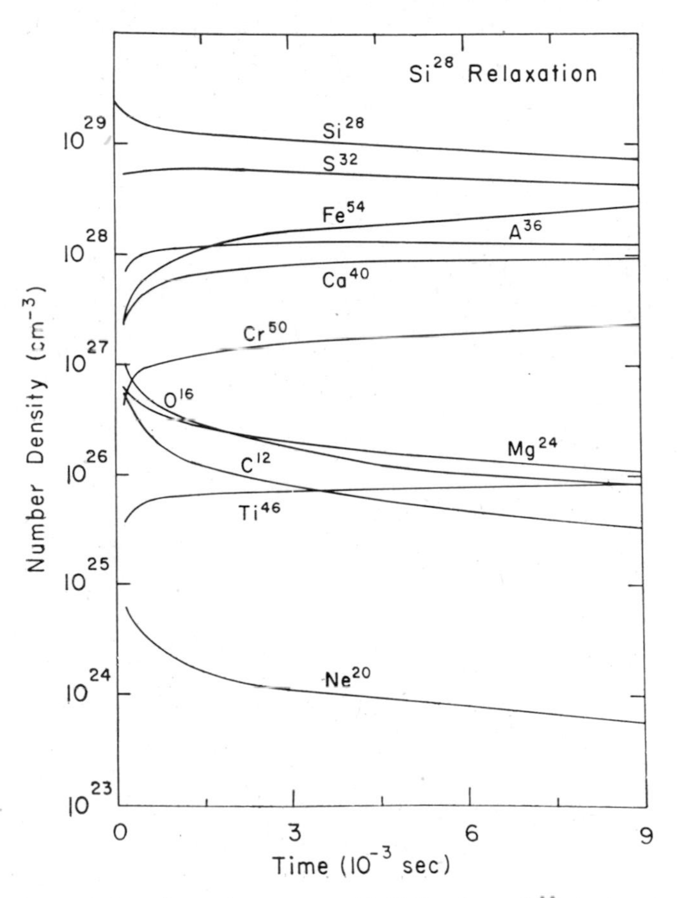

Fig. 3. The evolution of a region composed initially of pure Si^{28} at a temperature $T_9 = 3$.

abundances tend to decrease at a rate comparable with the rate of decrease of silicon. Our calculations were carried to the point at which computer roundoff errors rendered further results unreliable at about the one percent level of uncertainty. In both cases, the time scales of decrease of Si^{28} are considerably longer than the nuclear photodisintegration times

$$(\lambda^{-1} \sim 1.4 \times 10^{-4}, T_9 = 5; \lambda^{-1} \sim 3 \times 10^3, T_9 = 3).$$

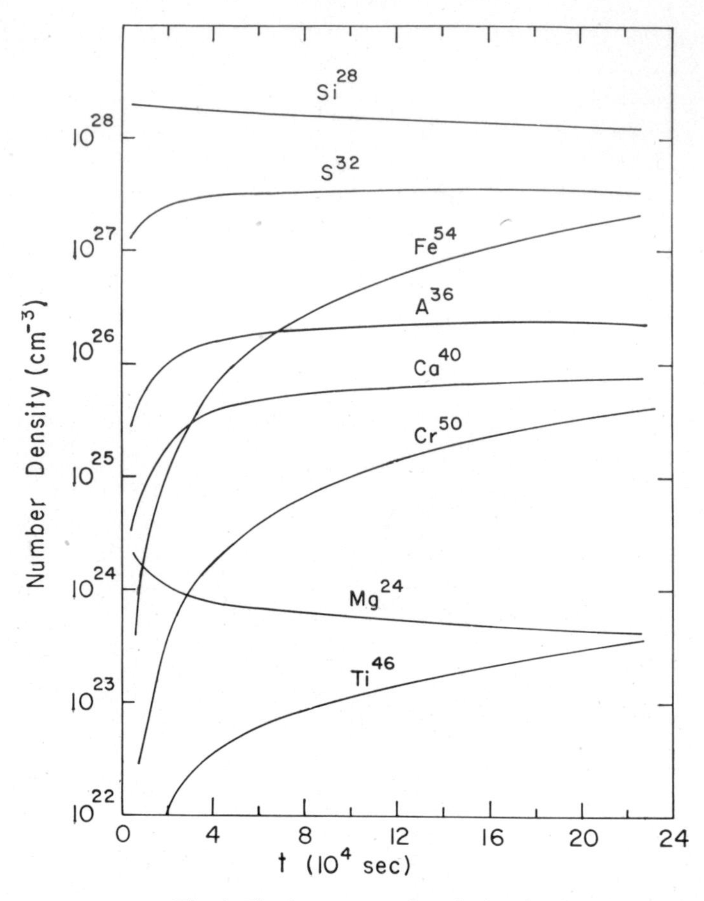

Fig. 4. The late stages of evolution for $T_9 = 5$.

The buildup of Mg^{24} and Al^{27} toward an effective equilibrium with Si^{28} results in a decrease in the net rate of destruction of silicon. That is, as the rate of Mg^{24} (α, γ) Si^{28} becomes comparable to that of Si^{28} (γ, α) Mg^{24} the net flow from silicon to Mg^{24} will decrease. A measure of the effective rate of destruction of silicon is the ratio of the net flow down the chain from silicon to the total photodisintegration rate of silicon. These ratios are plotted in Figures 5 and 6 as a function of the fraction of the initial silicon remaining in the medium. The ratio of the net flow to the total photodisintegration rate is seen to decrease as the abundance of silicon decreases in time. This is a measure of the degree to which the forward and inverse reaction rates have come into equilibrium. The smoothness of these curves suggests that

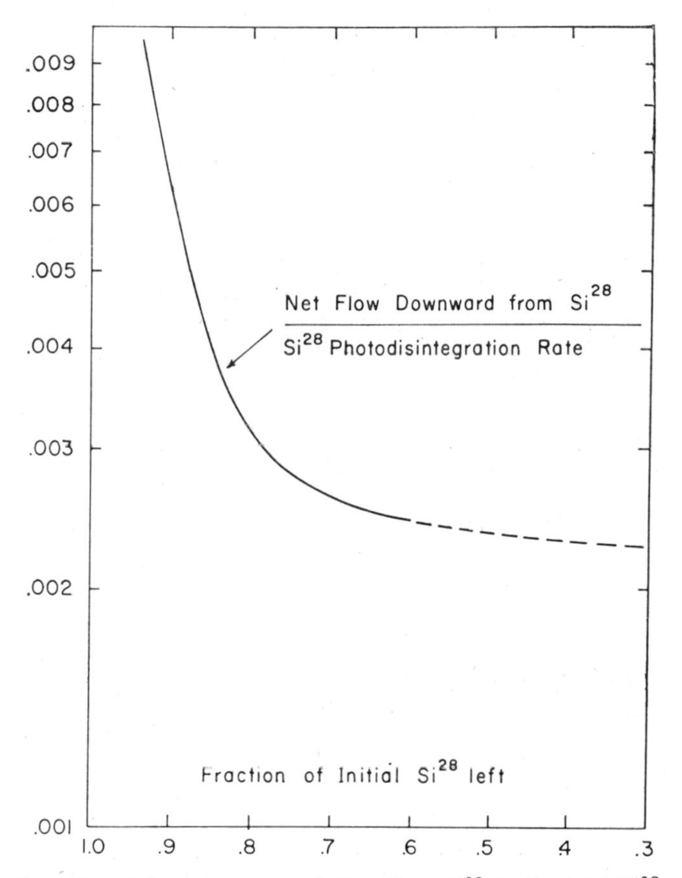

Fig. 5. The ratio of the net downward flow from Si^{28} to the total Si^{28} photodisintegration rate plotted as a function of the fraction of the initial Si^{28} remaining in the medium for $T_9 = 3$.

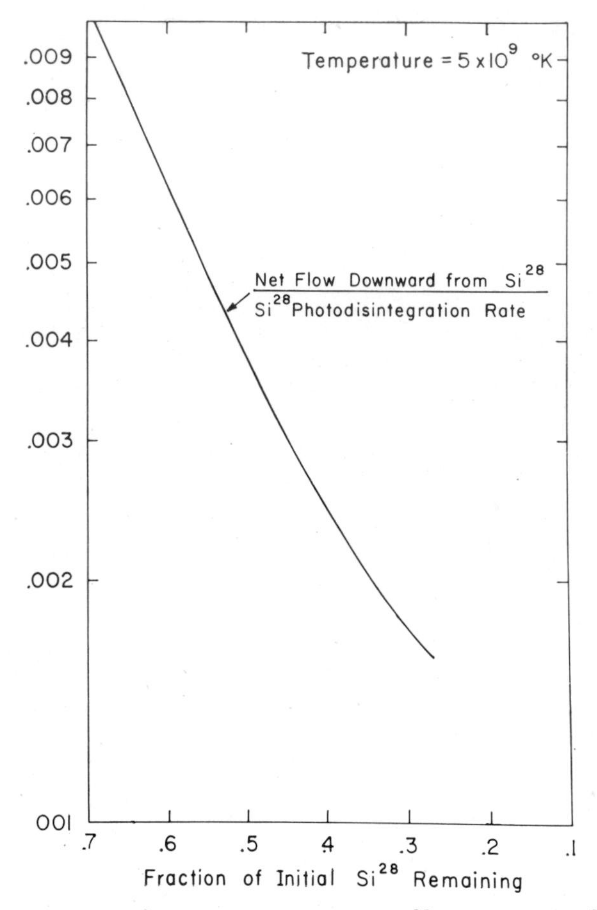

Fig. 6. The ratio of the net downward flot from Si^{28} to the total Si^{28} photodisintegration rate plotted as a function of the fraction of the initial Si^{28} remaining in the medium for $T_9 = 5$.

we might roughly predict the future behavior of the relaxation by an extrapolation of this dependence.

The dominant flows in the network are shown in Figure 7. Generally, they support the arguments presented in this discussion. The net flow down the chain from Si^{28} to alpha-particles is evident in this picture. The flow from C^{12} to alpha-particles proceeds very slowly, due to the slow rate of the C^{12} $(\gamma, 3\alpha)$ reaction. An equilibrium between C^{12} and α-particles is not established on the time scale of this relaxation. Above Si^{28} the alpha-particle links contribute appreciably to the evolution through A^{36}. Beyond this point (α, p), (p, γ) and (p, n) reactions carry the major flow toward the iron region.

The abundance distributions as a function of mass number at the end of our relaxation calculations are shown in Figures 8 and 9. The general features of these distributions can be understood in terms of the nuclear reac-

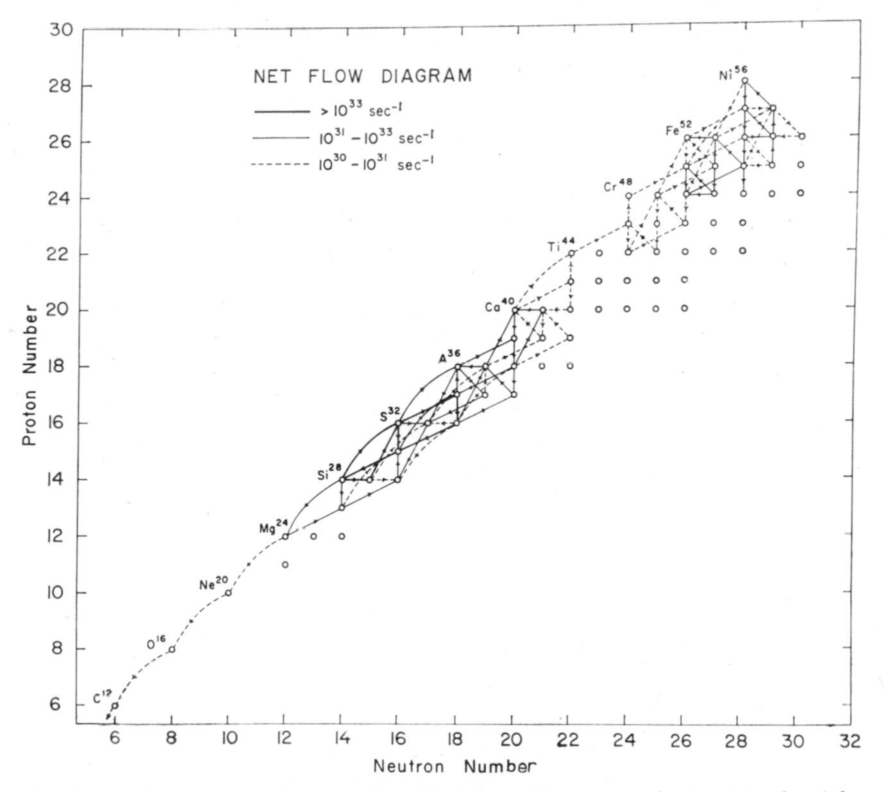

Fig. 7. The dominant net flows in the network are illustrated. The break in the alpha-particle chain past Ti^{44} is evident in this diagram.

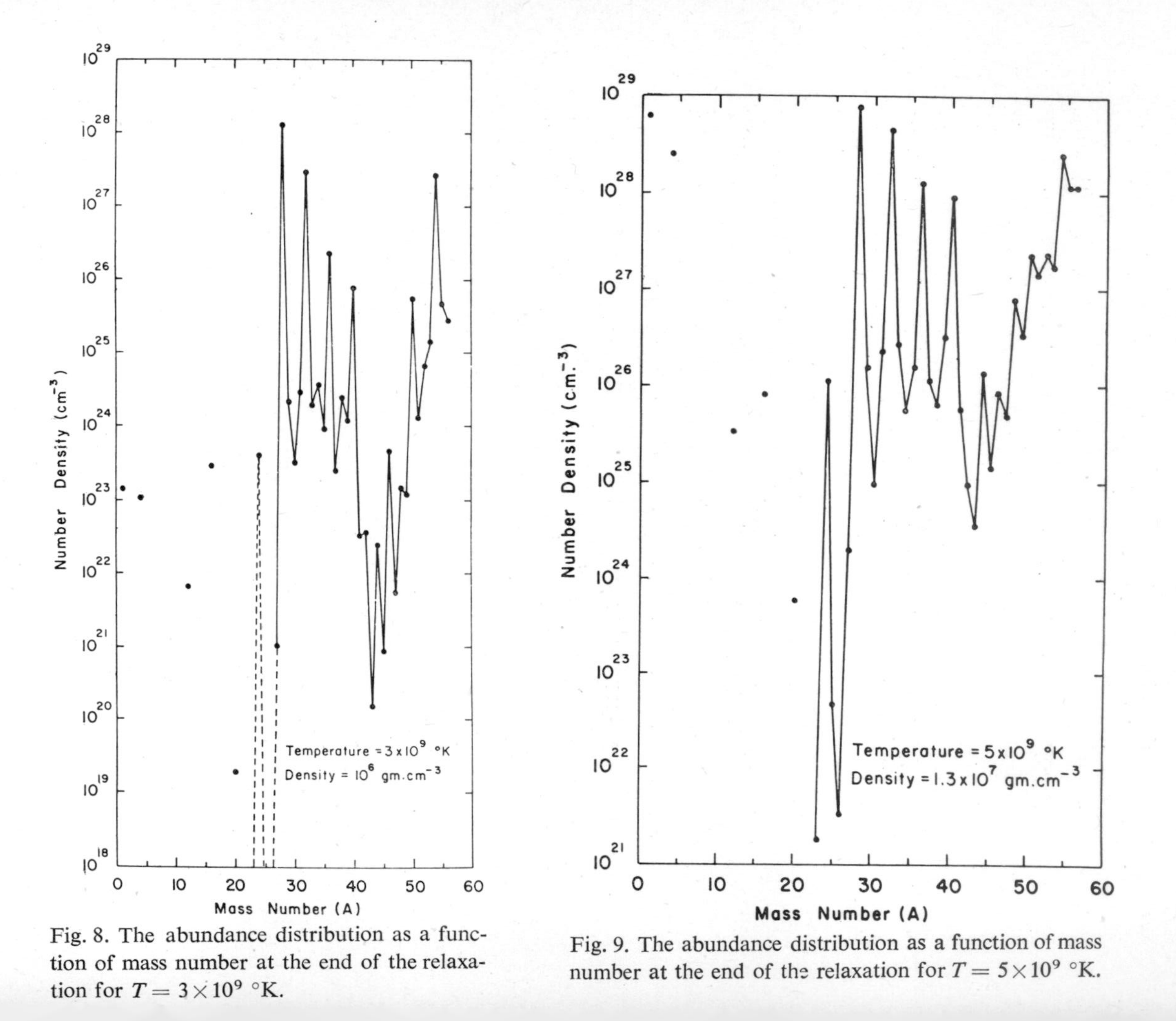

Fig. 8. The abundance distribution as a function of mass number at the end of the relaxation for $T = 3 \times 10^9$ °K.

Fig. 9. The abundance distribution as a function of mass number at the end of the relaxation for $T = 5 \times 10^9$ °K.

tion rate properties. The high abundances of protons and alpha-particles at $T = 5 \times 10^9$ °K are the result of the rapid photodisintegration rates at these temperatures. These have resulted, as well, in relatively larger abundances of the alpha-particle nuclei below silicon. The break in the alpha-particle chain past Ca^{40}, resulting in the production of the $\alpha + 2n$ nuclei Ti^{46}, Cr^{50} and Fe^{54} is evident in both distributions. However, the strong peaking at these mass numbers ($A = 46$, 50, and 54) at 3×10^9 °K is not apparent at 5×10^9 °K, due to the fact that at the higher temperatures many more reaction links are providing significant contributions to the flows.

The results of these calculations have important implications regarding nucleosynthesis.

1) The governing rate for the silicon to iron conversion is the effective loss rate of Si^{28}. This means that the time required for the silicon-to-iron conversion is not a sensitive function of the density. The buildup of nuclei in the vicinity of Si^{28} results in the establishment of flows in opposition to the downward flows from silicon. As these abundances approach an equilibrium, the net flows out of Si^{28} will decrease; hence the net rate of depletion of Si^{28} will decrease. The ratios of the net flow downward from Si^{28} to the total Si^{28} photodisintegration rate (the sum of the proton and alpha-particle photodisintegration rates) are shown in Figures 5 and 6 for the temperatures $T_9 = 3$ and 5 respectively.

These figures differ somewhat due to the different character of the flows at the two temperatures. A reasonable generalization of these two cases is that over a rather large range in the silicon abundance the ratio is typically 10^{-3} to 10^{-2}. This means that the time required for the Si^{28} abundance to decrease to one half its initial value is roughly 100 to 1000 times the photodisintegration half-life. This generalization should allow for a crude estimate of the time scale for the conversion of Si^{28} to iron at a different temperature. The total Si^{28} photodisintegration rate, taken to be the sum of the rates for (γ, p) and (γ, α) reactions, is given as a function of temperature in the range $1 \leqq T_9 \leqq 9$ by the approximate expression (Truran et al., 1966):

$$\begin{aligned} \lambda &= \lambda_p + \lambda_\alpha \\ &= 1.13 \times 10^{10} \exp\left(18.18 - \frac{7.85}{T_9^{1/3}} - \frac{139.45}{T_9}\right) \\ &\quad + 7.54 \times 10^{10} \exp\left(21.68 - \frac{21.77}{T_9^{1/3}} - \frac{121.73}{T_9}\right) \end{aligned}$$

2) Above Ca^{40}, the most abundant nuclei (Ti^{46}, Cr^{50}, Fe^{54}) do not have equal numbers of neutrons and protons. This is not because of beta decay: the contributions from this process are negligible, simply because beta-unstable nuclei do not build up to high enough abundances during the time scale of the relaxations considered here. It is, rather, a question of which reactions are energetically favorable.

Fowler and Hoyle (1964) have suggested a crude model in which Si^{28} photodisintegrates by (γ, α) and captures the α particles to build nuclei with $N = Z$ up to Ni^{56}. The present results suggest this model is unrealistic; for instance, at $T_9 = 3$, Fe^{54} is more abundant by about two orders of magnitude than any other nucleus with $A > 52$. This becomes even more clear when the actual reaction rates across the various links in the network are considered, as has been illustrated in the diagram of the net flows (Figure 7). Above Ca^{40} it was found that the alpha-particle links did not carry the major flows. This can be explained by the rapid decrease of alpha-penetrabilities with Z at a given energy.

3) Fe^{54} is the most abundant nucleus produced in the iron region. In the virtual absence of beta decays, this means that the overall reaction is not $2\ Si^{28} \rightarrow Ni^{56}$, which releases about 10 MeV per reaction, but $2\ Si^{28} \rightarrow Fe^{54} + 2p$, which is actually endoenergetic by 1.3 MeV. The implication is that there is no exoenergetic reaction phase; nuclear reactions involved in the approach to equilibrium do not slow down that collapse of a star brought on by neutrino emission. The exact energy release depends some what on the composition of the material in the iron region, but if Fe^{54} is the most abundant nucleus, the Si-Fe conversion can not be exothermic by much, if indeed at all. Nor will beta decays help the situation. It is true that if an electron or a positron is emitted it carries some of the energy release of the beta decay. However, there is also electron capture, which becomes important at high density. In fact, even normally stable nuclei can capture electrons when the electron Fermi energy becomes high. Fowler and Hoyle (1964) give a half-life of 4×10^4 sec. for electron capture at a temperature of 3.8×10^9 °K and a density of 3.1×10^6 gm cm^{-3}. In this kind of reaction, the entire energy release goes into neutrinos.

4) It is evident that the iron peak observed in nature, centered on Fe^{56}, cannot result from our calculations. However, the transformation of protons to neutrons at an early stage of this relaxation would result in the production of increased amounts of neutron rich isotopes of the elements in the iron region.

Chiu (1966) has studied the presupernova stage of evolution of a star in some detail. He finds that stars in this stage will evolve to a high central density ($\varrho \sim 10^{9-10}$ gm cm^{-3}) at a central temperature $T \sim 3 - 4 \times 10^9$ °K. At these densities, electron capture on nuclei will proceed rapidly. In particular, the electron capture rates for S^{32} and P^{32} are $> 10^{-1}$ sec.$^{-1}$. In times of the order of 100 seconds, an appreciable amount of S^{32} can be converted through P^{32} to Si^{32}. The processing of a region composed of Si^{28} and substantial amount of Si^{32} by a shock wave of peak temperature $\sim 5 \times 10^9$ °K might result in the formation of an iron abundance peak centered on Fe^{56}.

5) The relaxation of 5×10^9 °K resulted in the production of large amounts of iron peak nuclei ($\gtrsim 20\%$ by mass) in $\sim 10^{-2}$ seconds. This is approximately the time for which the post shock wave temperature predicted by the model of Colgate *et al.* (1961) will remain $\gtrsim 5 \times 10^9$ °K. It is evident, therefore, that significant amounts of iron can be produced by the passage of the shock through a silicon region.

REFERENCES

Burbidge, E. M., Burbidge, G. R., Fowler, W. A. and Hoyle, F. (1957). *Rev. Mod. Phys.* **29**, 547.

Cameron, A. G. W. (1959a). *Astrophys. J.* **130**, 429.

(1959b). *Astrophys. J.* **130**, 895.

(1963). Nuclear Astrophysics, compilation of lectures given at Yale University.

Chiu, H. Y. (1966). *Presupernova Evolution, in Stellar Evolution*, R. F. Stein and A. G. W. Cameron (Eds.), Plenum Press, N.Y., in press.

Colgate, A., Grasberger, W. H. and White, R. H. (1961). *The Dynamics of a Supernova Explosion*, Lawrence Radiation Laboratory UCRL-6471.

Fowler, W. A. and Hoyle, F. (1964). *Astrophys. J. Suppl.* **91**, 1.

Hayashi, C., Hoshi, R. and Sugimoto, D. (1962). *Suppl. Prog. Theor. Phys.* **22**, 1.

Stothers, R. and Chiu, H. Y. (1962). *Astrophys. J.* **135**, 963.

Truran, J. W., Cameron, A. G. W. and Gilbert, A. (1966). *Can. J. Phys.* (**44**, 563.)

Note added in proof: The emergence of Fe^{54} as the most abundant nucleus in the iron peak in the relaxation calculation performed at 3×10^9 °K implies that our results have been influenced to some extent by beta-decay. For a pure N = Z mixture, the equations of statistical equilibrium predict that Ni^{56} will be the most abundant nucleus under these conditions ($T_9 = 3$, $\varrho = 10^6$ gm/cc). The same roundoff errors that forced us to terminate our calculation render any analysis of the total beta-decay contribution impossible; i.e., for the entire system $(N\text{-}Z)/Z < 0.01$. Subsequent improved calculations have revealed that this neutron enrichment is indeed realized.

Equilibrium Composition of Matter at High Densities

SACHIKO TSURUTA

We have seen that in the later stages of stellar evolution, such as the pre-implosion phase of a supernova, central temperatures and densities approach very high values. Under such conditions the processes of nucleosynthesis yield exotic mixtures of nuclei which may be reflected in the observed abundance of the elements. Another environment for such conditions may exist in the surface layers of neutron stars where x-ray emission might be detectable in some future experiment.

For the calculations to be reported here, I will restrict the ranges of temperature and density to lie within 2–10 billion °K and electron Fermi levels of 0.17–23 MeV., corresponding to densities of roughtly 10^6–10^{11} gm/cm^3 depending on the detailed composition of the mixture.

The equilibrium part of the calculation will rest on the two assumptions that nuclear statistical equilibrium (NSE) holds and that the system is in a steady state condition with respect to weak interactions. By the latter we mean that the total rates of neutrino and antineutrino emission are equal, thus implying that there is no change in the total neutron or proton number with time. The assumption of NSE is the main reason why the temperature and density ranges were so chosen. At the lower end of the range the rates of nuclear processes are slowed down to the extent that the necessary conditions for NSE are no longer obtainable, while at the high end of the range all nuclei are broken down by photodisintegration and electron capture reactions into neutrons.

The basic equation for nuclear abundances in NSE is given by an expression of the form (Burbidge, Burbidge, Fowler and Hoyle 1957—B^{2}FH)

$$n(A, Z) = f(A, Z, T) \{n_n^{A-Z} n_p^Z\}$$

where n(A, Z) is the number density of the nucleus with nucleon number A and charge number Z, n_n and n_p are the neutron and proton number densities, and T is the temperature. The function f(A, Z, T) contains all other necessary information such as the binding energy and nuclear level characteristics of nucleus (A, Z), (and hence its partition function), etc. Nuclear masses were obtained using Cameron's (1957) semi-empirical mass formula. Nuclear level densities were assumed to be of the form

$$N = \exp [a(U - U_0)],$$

where N is the total number of levels up to excitation energy U, and a and U_0 are parameters fit on the basis of level structures of known nuclei.

The neutrino processes were restricted to beta decay and electron capture. Because of the high temperatures involved a significant portion of beta decays for a given nucleus come from decays of excited states. The total decay rate per nucleus is then

$$P^- = \sum_i a_i (\ln 2) f_i^- / (ft)_i$$

where $(ft)_i$ is the terrestrial ft value for the transition, f_i is the fermi function for the transition which in this case includes inhibition factors that take care of limitations in available phase space due to the presence of degenerate electrons in the surrounding medium. The factor a_i is the fractional population of the ith level.

The inverse process of electron capture was calculated using a method due to Bahcall (1964) which gives the electron capture rate per nucleus in the form

$$\lambda = (\ln 2) f_c / (ft),$$

where f_c is related to the Fermi function and includes the effects of electron degeneracy. This rate is to be summed over all contributing levels to yield the total capture rate P^+. For weak interaction steady state we must have

$$\sum_i P_i^- n^- (A_i, Z_i) = \sum_k P_k^+ n^+ (A_k, Z_k).$$

For a given temperature and density an iterative procedure was used to determine the equilibrium number densities. These are shown in the accompanying set of figures, the first six of which are at a typical temperature of 5×10^9 °K.

Figure 1 shows the composition when the density is 3.86×10^5 gm/cm^3. The solid lines are even charged nuclei and the dashed lines odd charged. As is consistent with the e-process of B^2FH and the work of Cameron, the most abundant nucleus is Fe^{56} with neighboring nuclei decreasing in abundance as one moves away from Fe^{56}.

The case for a density of 2×10^9 gm/cm^3 is shown in Figure 2 where the most abundant nucleus has shifted to Ni^{66}. The effects of further increases

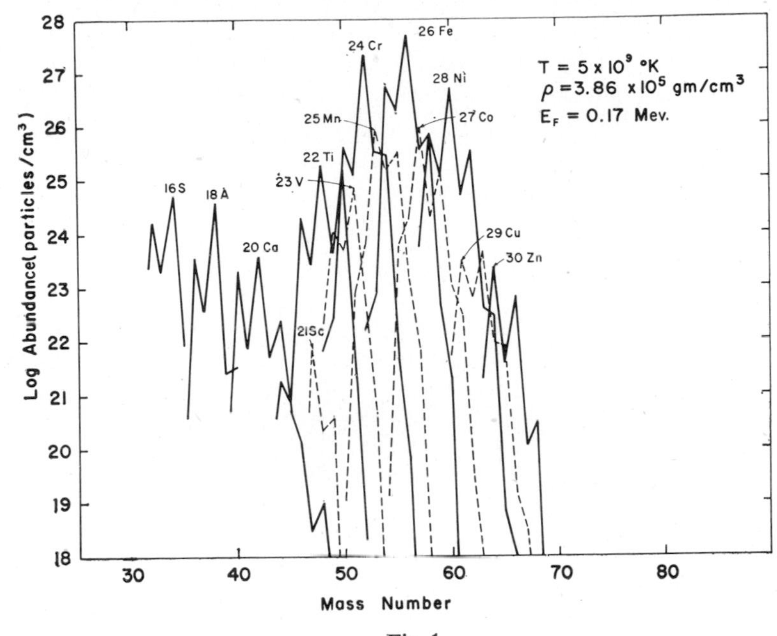
T = 5 x 10⁹ °K
ρ = 3.86 x 10⁵ gm/cm³
E_F = 0.17 Mev.
Log Abundance (particles/cm³)
Mass Number
16 S
18 A
20 Ca
21 Sc
22 Ti
23 V
24 Cr
25 Mn
26 Fe
27 Co
28 Ni
29 Cu
30 Zn
28
27
26
25
24
23
22
21
20
19
18
30
40
50
60
70
80

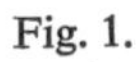
Fig. 1.

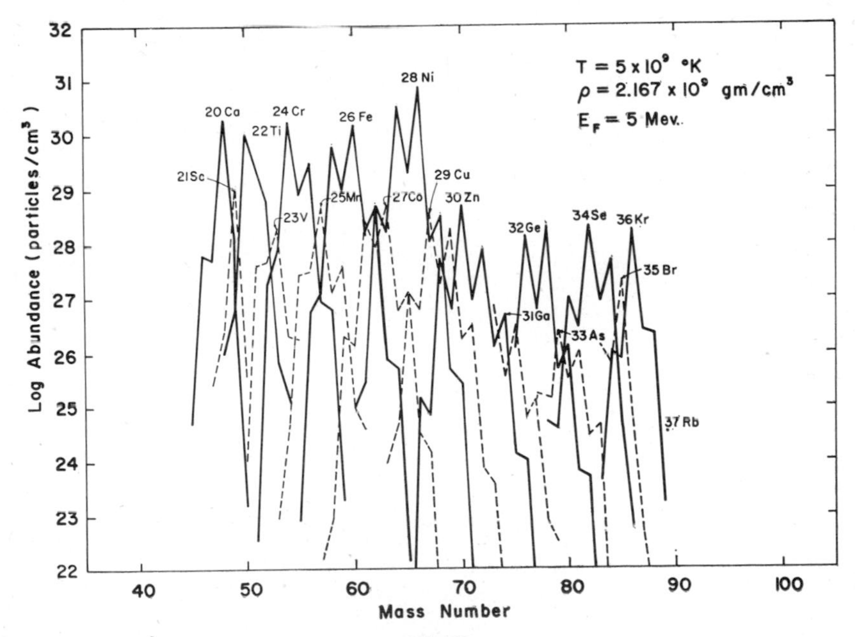
T = 5 x 10⁹ °K
ρ = 2.167 x 10⁹ gm/cm³
E_F = 5 Mev.
Log Abundance (particles/cm³)
Mass Number
20 Ca
21 Sc
22 Ti
23 V
24 Cr
25 Mn
26 Fe
27 Co
28 Ni
29 Cu
30 Zn
31 Ga
32 Ge
33 As
34 Se
35 Br
36 Kr
37 Rb
32
31
30
29
28
27
26
25
24
23
22
40
50
60
70
80
90
100

Fig. 2.

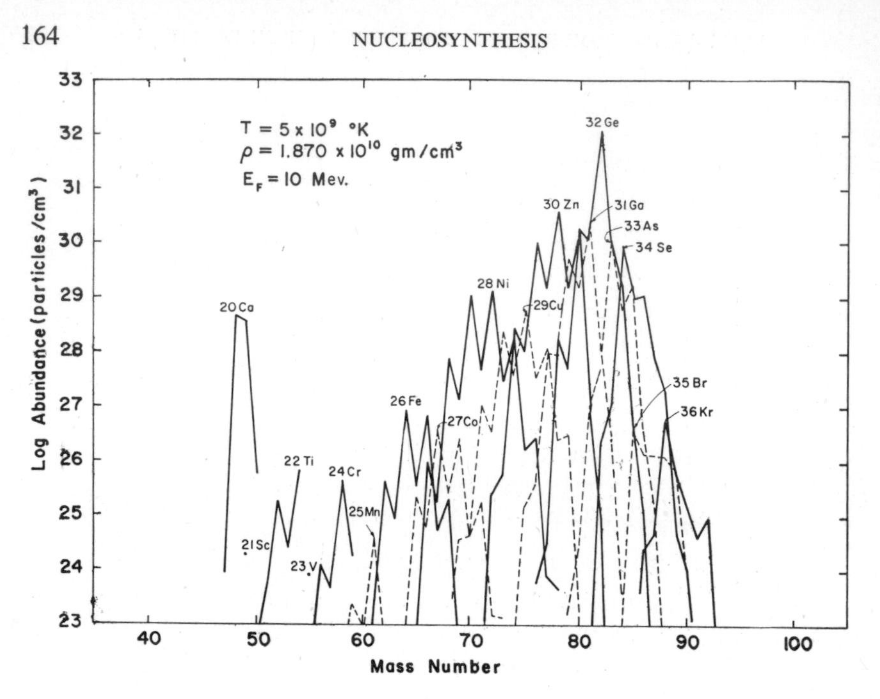
T = 5 x 10⁹ °K
ρ = 1.870 x 10¹⁰ gm/cm³
E_F = 10 Mev.
Log Abundance (particles/cm³)
Mass Number
20 Ca
21 Sc
22 Ti
23 V
24 Cr
25 Mn
26 Fe
27 Co
28 Ni
29 Cu
30 Zn
31 Ga
32 Ge
33 As
34 Se
35 Br
36 Kr
23
24
25
26
27
28
29
30
31
32
33
40
50
60
70
80
90
100

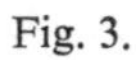

Fig. 3.

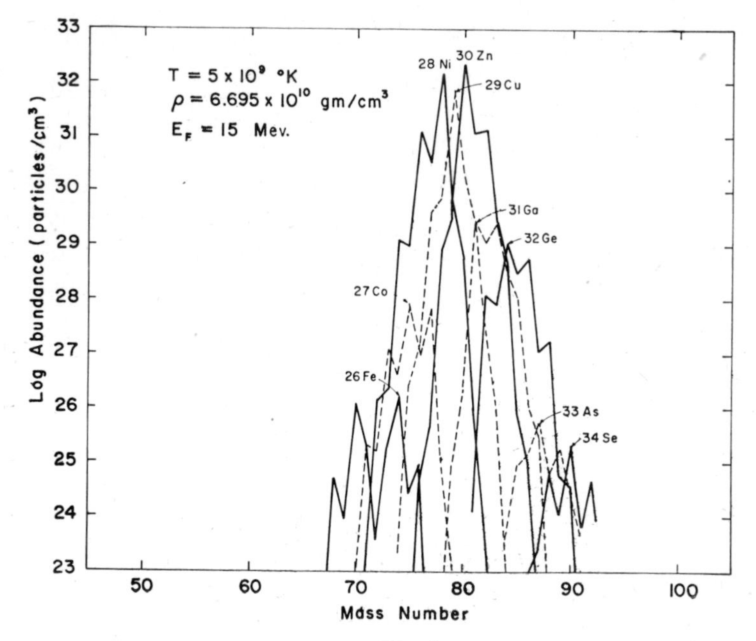
T = 5 x 10⁹ °K
ρ = 6.695 x 10¹⁰ gm/cm³
E_F = 15 Mev.
Log Abundance (particles/cm³)
Mass Number
26 Fe
27 Co
28 Ni
29 Cu
30 Zn
31 Ga
32 Ge
33 As
34 Se
23
24
25
26
27
28
29
30
31
32
33
50
60
70
80
90
100

Fig. 4.

in density are shown in Figures 3–6 where in the final figure two distinct abundance peaks are visible, centered around the Fe^{78} – Ni^{84} region and around Sr^{120} which corresponds to neutron magic number 82. The trend, then, as density increases is a gradual shift in composition to neutron rich matter.

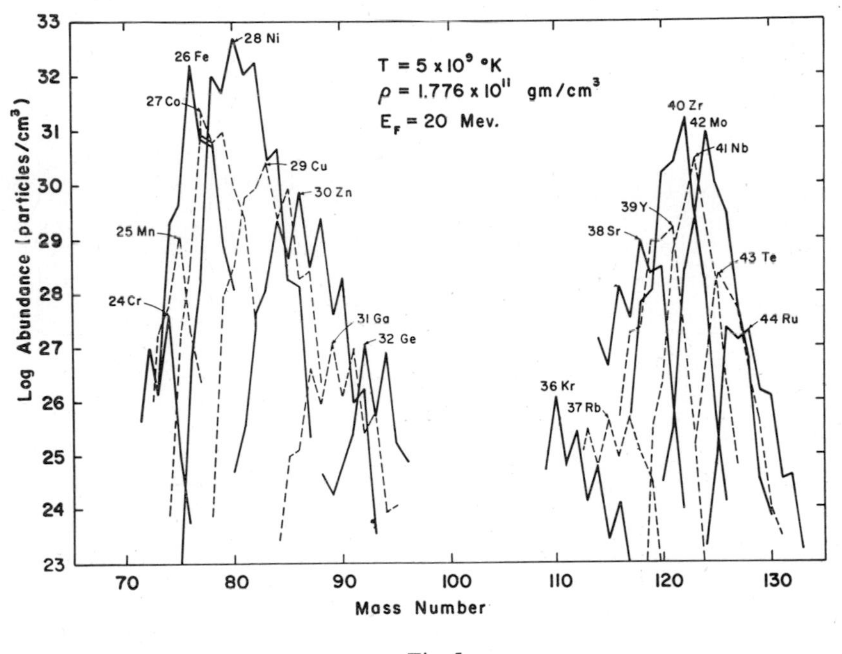

Fig. 5.

Figure 7 illustrates the effect of temperature (taken at 5×10^9 °K and 2×10^9 °K) with the density of Figure 3. The most abundant nucleus is still Ge^{82}, but the abundances of neighboring nuclei have dropped with decreased temperature to the point where all other constituents are negligible. Another general effect of temperature is the enhancement of less neutron rich nuclei as temperature increases. When the temperature exceeds 10^{10} °K, regardless of density, the material breaks down into neutrons.

In Figure 8 is shown the URCA loss rates associated with the sum of beta decay and electron capture. At a typical temperature of 5×10^9 °K a density increase of from 10^6 to 3×10^{11} gm/cm^3 increases the loss rate from about 10^{11} erg/gm-sec to 10^{17} erg/gm-sec.

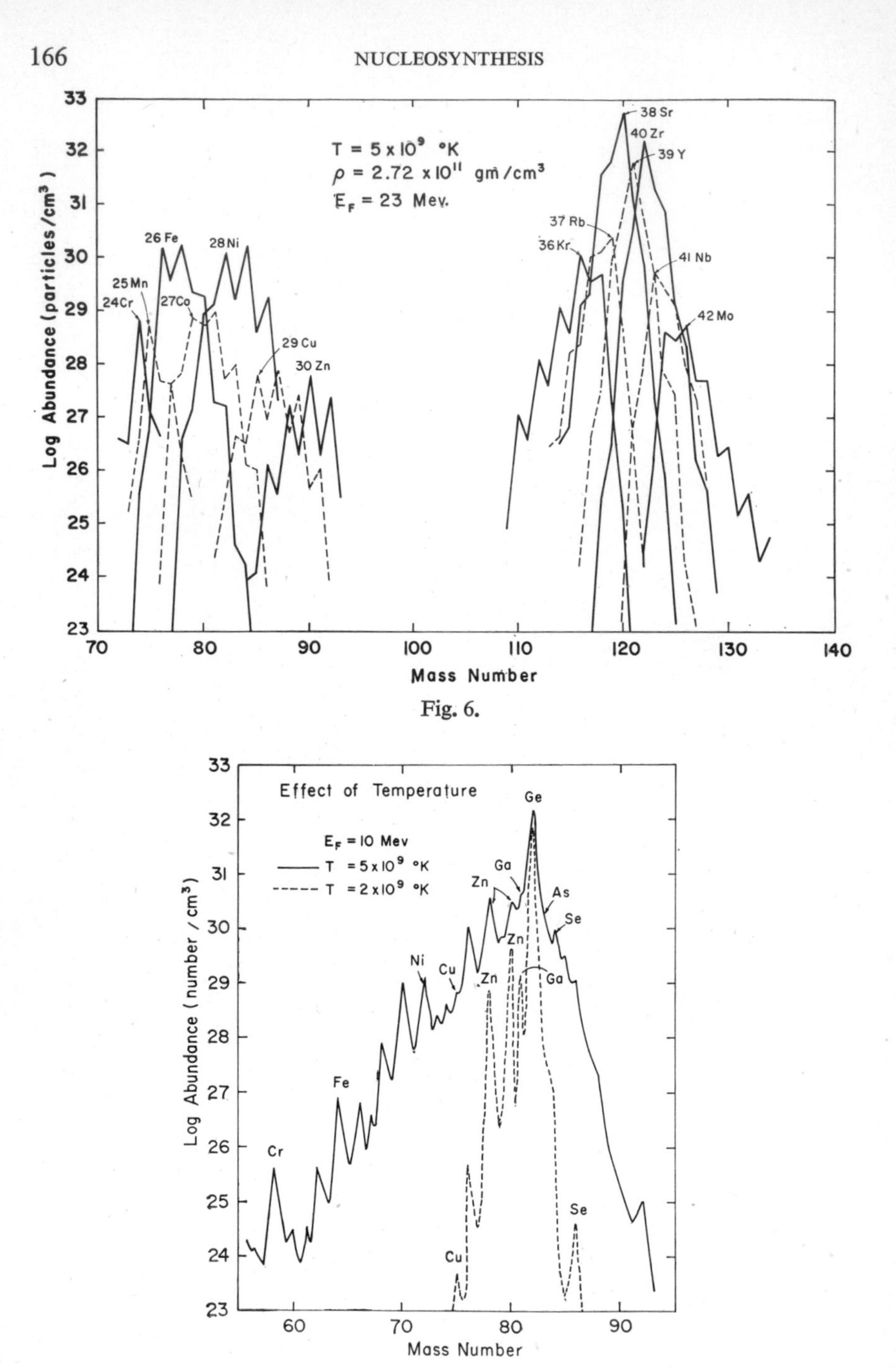
T = 5 x 10⁹ °K
ρ = 2.72 x 10¹¹ gm/cm³
E_F = 23 Mev.
Log Abundance (particles/cm³)
Mass Number
24Cr
25Mn
26Fe
27Co
28Ni
29Cu
30Zn
36Kr
37Rb
38Sr
39Y
40Zr
41Nb
42Mo
Effect of Temperature
E_F = 10 Mev
T = 5 x 10⁹ °K
T = 2 x 10⁹ °K
Log Abundance (number/cm³)
Cr
Fe
Ni
Cu
Zn
Ga
Ge
As
Se

Fig. 6.

Fig. 7.

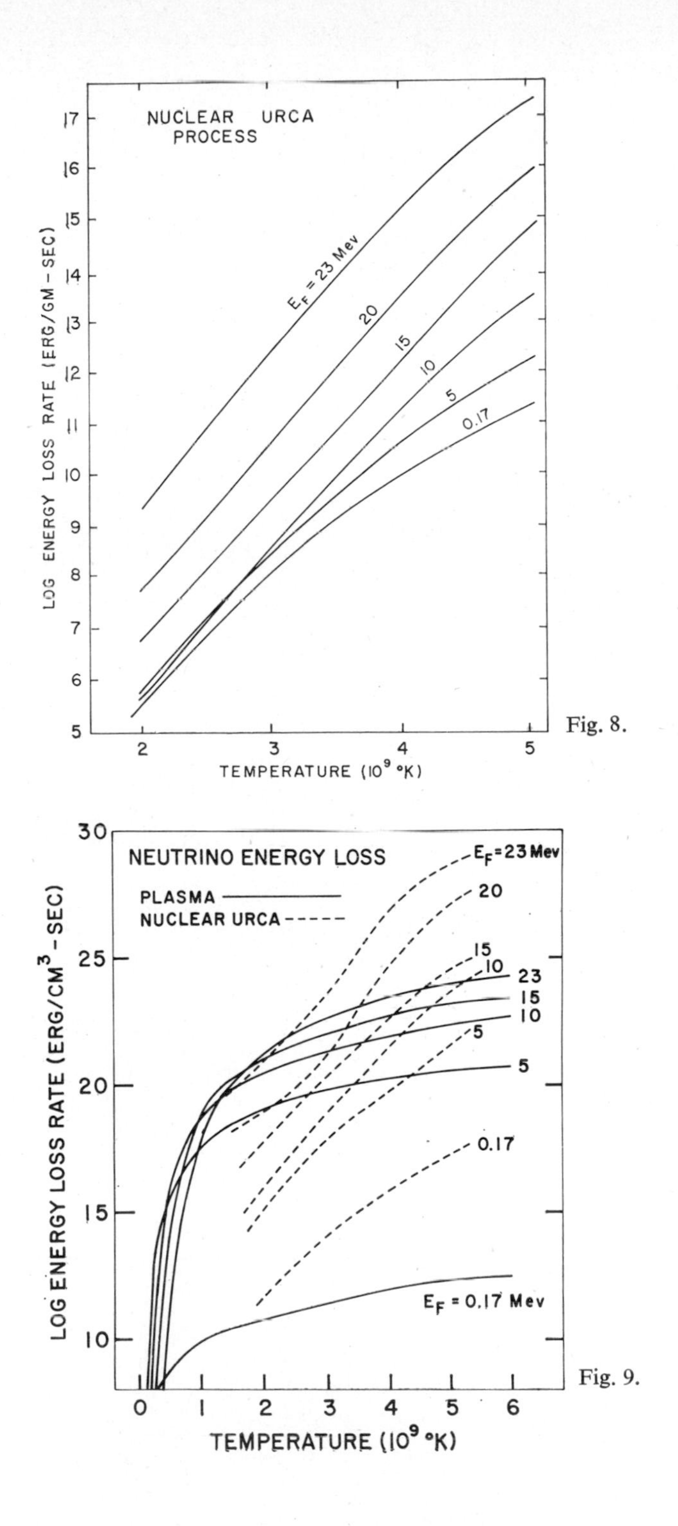
NUCLEAR URCA PROCESS
LOG ENERGY LOSS RATE (ERG/GM - SEC)
TEMPERATURE (10^9 °K)
E_F = 23 Mev
20
15
10
5
0.17
NEUTRINO ENERGY LOSS
PLASMA
NUCLEAR URCA
LOG ENERGY LOSS RATE (ERG/CM^3 - SEC)
TEMPERATURE (10^9 °K)
E_F = 23 Mev
20
15
10
23
15
10
5
5
0.17
E_F = 0.17 Mev

Fig. 8.

Fig. 9.

A comparison of the URCA rate to the plasmon neutrino loss rate (Inman and Ruderman 1964) is shown in Figure 9 where dashed curves refer to URCA and solid curves to plasmon losses. For the lower range of temperature plasmon neutrinos dominate, but at high temperature extreme the URCA process takes over.

REFERENCES

Bahcall, J. N. (1964). *Astrophys. J.* **139**, 318.

Burbidge, E. M. Burbidge, G. R., Fowler, W. A. and Hoyle, F. (1957). *Rev. Mod. Phys.* **29**, 547.

Cameron, A. G. W. (1957), *Can. J. Phys.* **35**, 1021.

Inman, C. L. and Ruderman, M. A. (1964). *Astrophys. J.* **140**, 1025.

Tsuruta, S. *Neutron Star Models*, Thesis, Columbia University, (1964) unpublished.

Tsuruta, S., and Cameron, A. G. W. (1965). *Can. J. Phys.* **43**, 2056.

DISCUSSION

S. A. Colgate: The neutrino loss rates calculated here for the equilibrium case differ significantly from the rates that would be calculated for the approach to NSE on a fast time scale.

A. G. W. Cameron: Yes, the case given here would obtain from the approach to NSE if the system were allowed a few seconds to relax.

The Nuclear Synthesis Conditions of Types I and II Supernovae

S. A. Colgate

In order to provide the theoretical background for discussion of nuclear synthesis conditions in supernova, we begin with a brief summary of the supernova process.[1]

EVOLUTION OF A SUPERNOVA INTERIOR

The neutral stability line for equilibrium support of a star against its gravitational attraction is the path along which the star can be compressed or expanded, doing no work. Figure 1 shows this line in the pressure-density plane for a 10 $M_{\odot}$ polytrope of index 3; the slope corresponds to an adiabatic exponent $\gamma = 4/3$. Depending upon the mass of the star, the pressure required for support at a given density will be given by lines of this same slope. The minimum pressure possible at lower densities is due to electron degeneracy.

The solid heavy curve (in Figure 1) corresponds to the approximate adiabat followed by a 10 $M_{\odot}$ star during iron to helium conversion. The deficiency in pressure along this adiabat corresponds to a lack of support needed to maintain the star in hydrostatic equilibrium. Because this deficiency is so large, the matter is almost in free-fall. The stellar structure discussed earlier in this conference by Chiu refers to a cooler star whose support is electron degeneracy pressure. The central part of such a star is stable until the density rises high enough for electrons to combine with protons by inverse beta-decay, giving a pressure defect and forming neutron-rich matter. The number of leptons per baryon decreases, so the importance of the degeneracy pressure of the leptons decreases; this corresponds to the pressure defect in the region of density $\varrho \sim$ several . 10^{11} gm/cm^3.

[1] See Colgate and White (1964), for a more detailed treatment.

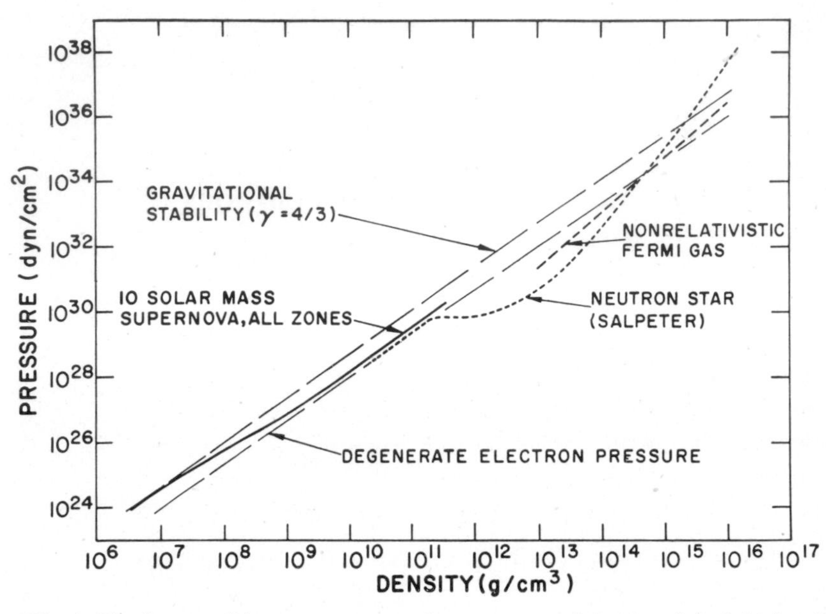

Fig. 1. The heavy solid curve represents the pressure and density of the initial and imploding zones of the supernova star whith no hard core. The dotted curve represents the cold neutron star equation of state of Salpeter with the degenerate electron and nucleon pressure curves included. It is expected that the final bounce will occur when the neutron star pressure equals the gravitational stability pressure at 2×10^{15} g/cm^3.

Regardless of whether or not the star becomes unstable through iron-helium conversion, it will become unstable in this higher density region by inverse beta-decay. The reason for emphasizing this point is that the rate of energy loss due to the URCA process discussed in the paper by Chiu[2] is based upon the assumption that the weak interactions are at equilibrium in this density region. If the matter is above equilibrium—as with a neutron-proton ratio of roughly unity—then in approaching equilibrium most of the inverse beta-decays occur on the free protons, which means that the energy carried away by the neutrino is of the order of the electron Fermi energy. In other words, thermal energy frees the proton, taking roughly 15 or 20 mev to do it. Then in the inverse beta-decay the neutrino has an energy roughly equal to that supplied by the electron, which is near the electron Fermi energy. This is why this cooling process is more effective as matter approaches equilibrium than when it is at equilibrium.

[2] See the paper presented by H. Y. Chiu at this conference.

The hydrostatic stability of the dense core depends upon the equation of state for dense neutron matter. Roughly speaking, the collapse of the core is halted when the pressure due to degenerate neutron matter for the given density rises above the equilibrium support curve. This "bounce point" corresponds to very high density. In Figure 1 there are two curves in this density region; one for a nonrelativistic gas of non-interacting nucleons, and a steeper curve which includes a nucleon hard-core potential term. The fraction of nucleon hard-core potential which is used in this equation of state is small in the region of bounce when the core contains only a few solar masses of matter.

Figure 2 shows the position of surfaces containing selected mass-fractions of the star as a function of time. The zero of the time-scale is arbitrary, corresponding to some point at which the calculation was begun. Note that the compression ratio becomes more than 10^3 and the core of the star forms with the neutron star equation of state. After this small stellar core is formed, he matter falling on its surface is suddenly stopped. When matter with a

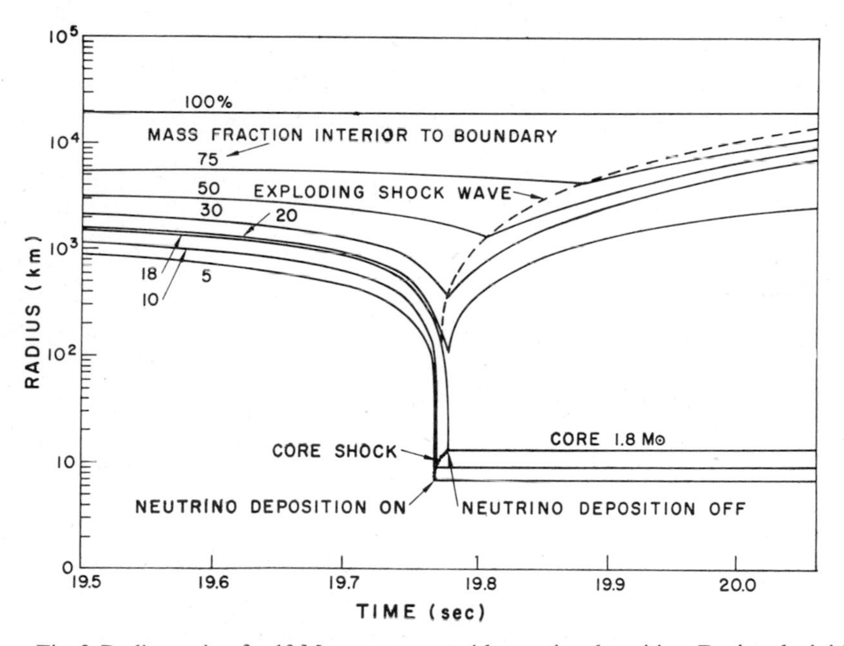

Fig. 2. Radius *vs* time for 10 M⊙ supernova with neutrino deposition. During the initial collapse the neutrino energy is assumed lost from the star, but at the time of formation of a core shock wave (heavy dots) a fraction of the neutrino energy is deposited in the envelope. The deposition ceases when the explosion terminates the imploding shock wave on the core.

large velocity and low thermal energy is suddenly stopped, a shock wave is formed. The internal energy behind the shock is just equal to the kinetic energy of the free-falling matter. The gravitational potential of a core of approximately one solar mass and radius of ten kilometers is roughly 100 MeV per nucleon. The temperatures behind the shock then correspond to 50 MeV per particle.

At such temperatures the neutrino opacity of this matter is high, with mean free paths of the order of tens of meters. Complete neutrino equilibrium is approached rapidly compared to the hydrodynamic time scale. The muon-type neutrino emission is negligible within the time of cooling by ordinary electrontype neutrinos, so that the concern here is heat diffusion carried by electron-type neutrinos out through the imploding material of the star. The calculations reported here were performed by taking the thermal energy created by the shock wave and distributing one-half of it throughout the rest of the matter of the star on the basis of the number of grams per square centimeter that the neutrino flux might traverse.

The matter is opaque to this neutrino flux inside a radius of roughly 100 kilometers. This may be derived in the following way. The rate at which energy is being generated in the shock wave is just the gravitational energy released while the shock is accumulating, divided by the time it takes to accumulate ($\sim$ 0.02 seconds). So

$$\text{Rate} \sim \frac{G(\Delta M)^2}{R\,\Delta t} \sim \frac{G(1\,M_\odot)^2}{R\ 0.02\ \text{sec}}$$

This is equated to the area of the emitting surface $4\pi R^2$ times the emission rate for a fermion black body, $ac/3\,T^4$. There is actually a 7/8 factor for zero-mass fermions instead of bosons. The surface appears to be roughly at a radius $R \sim 10^2$ km and a temperature T a bit less than 10 MeV. At the density corresponding to this radius, $\varrho \sim 10^{11}$ gm/cm^3, the neutrino mean-free path is about one-third to one-half the radius of the star. This defines the neutrino luminosity surface; for larger densities the neutrinos diffuse outward.

Figure 3 is essentially the same as Figure 2, but the radii are shown on a linear rather than logarithmic scale. The fact that the curvature of the shock-wave path is upward indicates that the shock is increasing in velocity as it goes outward. This is the explosion shock which blows off the outer layers of the star. The core shock is at zero radius and therefore not visible on this graph. Because the shock is speeding up within a gravitational field the velocities of the innermost zones are not as high as would be estimated by taking the square root of the energy density. These velocities are of the order of only 10^9 cm/sec or less. This is considerably less than the free-fall velocity corresponding to the gravitational potential of their original position. This means that the matter is expelled at lower velocity than might be expected if gravity were neglected.

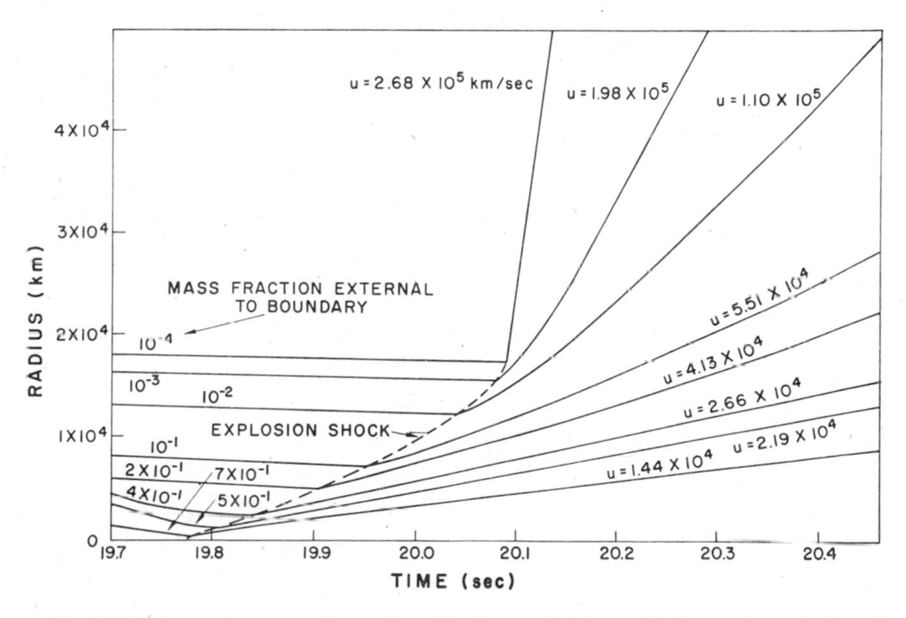

Fig. 3. 10 M⊙ supernova radius *vs* time. The linear plot shows the increasing velocity of the outward explosion shock wave reaching the relativistic limit for 10^{-4} mass fraction.

Figure 4 shows the same phenomena as Figure 2, but for a 2 M⊙ polytrope of index 3 stellar model. For such a small star, the evolution follows a low temperature path and the inverse beta-decay starts the collapse at a density $\varrho \sim 10^9$–10^{10} gm/cm³. Consequently, the initial radii are smaller than in the previous case of 10 M⊙ star. The time scale is expanded also. In this case the remainder of the star is ejected after a core of only 1 M⊙ is formed. The velocities are somewhat slower here, corresponding to the reduced gravitational potential of the core. Otherwise there is little difference. Figure 5 shows the linear plot of the history of a 2 M⊙ star. Again the explosion shock wave accelerates as it propogates outward. The bulk of the material has velocities of roughly 3×10^8 to 10^9 cm/sec. These velocities of expansion will be of interest in the latter part of this paper. We wish to know the adiabats of this matter, how high the temperature becomes when it is irra-

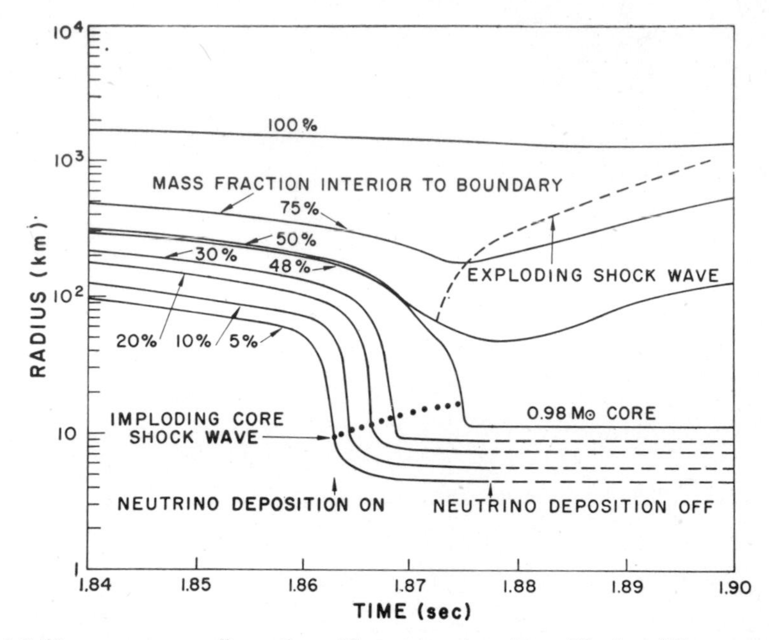

Fig. 4. 2 M⊙ supernova radius *vs* time with neutrino deposition. The instability occurs du to neutrino emission and nucleon binding in the equation of state with $\varrho > 2 \times 10^{11}\ \mathrm{g/cm^3}$.

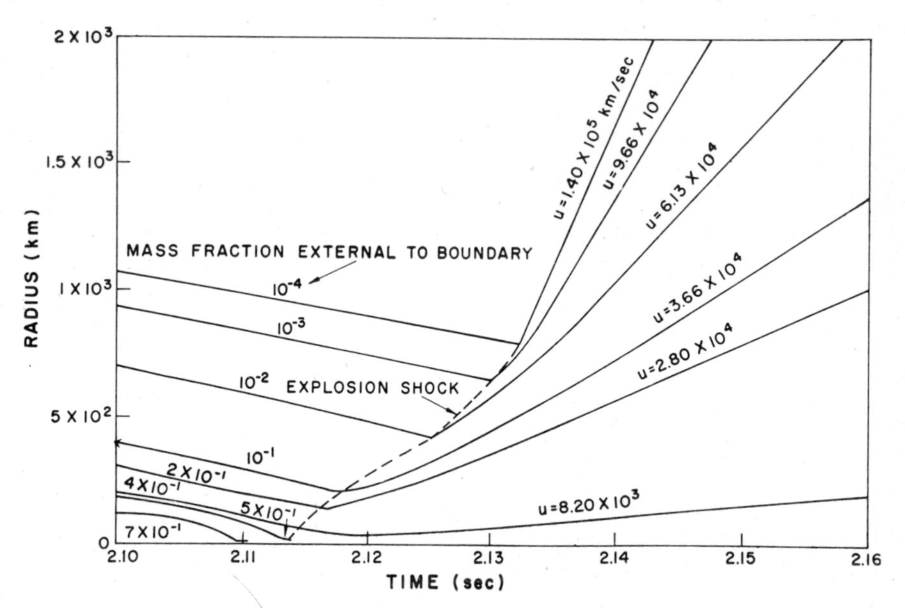

Fig. 5. 2 M⊙ supernova radius *vs* time with neutrino deposition.

diated by the neutrino flux, and the neutron-proton ratios during expansion and cooling.

Figures 6 and 7 show the history of a 1.5 M$_\odot$ star. In this case the time scale is further expanded. Otherwise it is similar to the 2.0 M$_\odot$ previously discussed.

The neutrino heat wave is terminated when

$$PV \sim \frac{MG}{R}$$

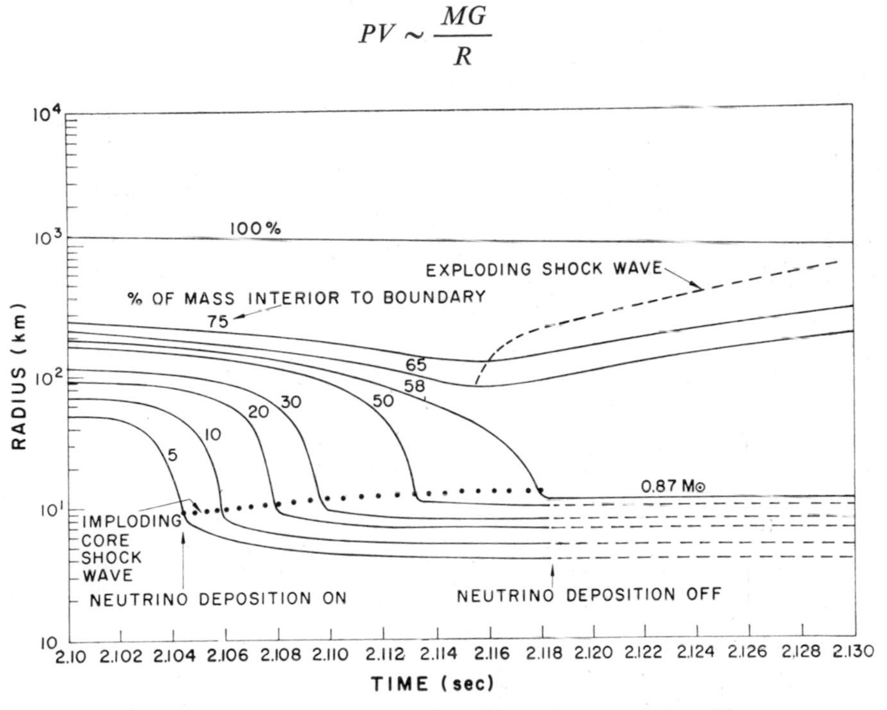

Fig. 6. 1.5 M$_\odot$ supernova radius *vs* time with neutrino deposition.

which occurs when the infalling zones reverse their velocity and begin to be blown off. Figure 8 illustrates the different time scales for expansion and compression. The zones which are blown off are first compressed to densities as high as $\varrho \sim 10^{11}$ gm/cm^3. Obtained from the equation of state and the equilibrium support curve, the pressure necessary to eject outlying matter implies a temperature of the order of 10 MeV. for a density of 10^{11} gm/cm^3. A small increase in temperature causes a large increase in pressure because the energy density is almost entirely due to radiation.

Figure 9 shows the adiabats followed by selected zones of the 10 M$_\odot$ during implosion and explosion. These calculations used an approximate

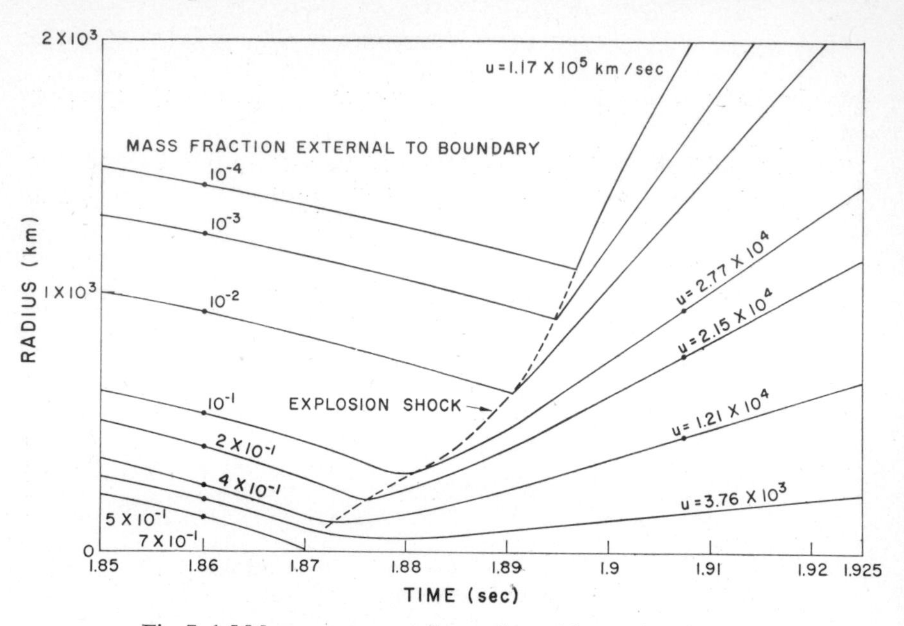

Fig. 7. 1.5 M$_\odot$ supernova radius *vs* time with neutrino deposition.

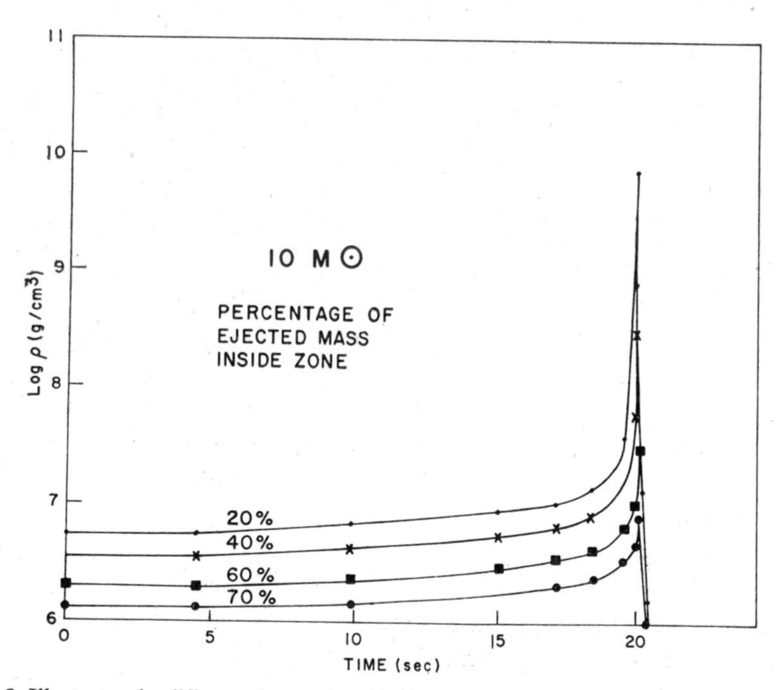

Fig. 8. Illustrates the different time scales for compression and expansion of various zones of the 10 M$_\odot$ problem.

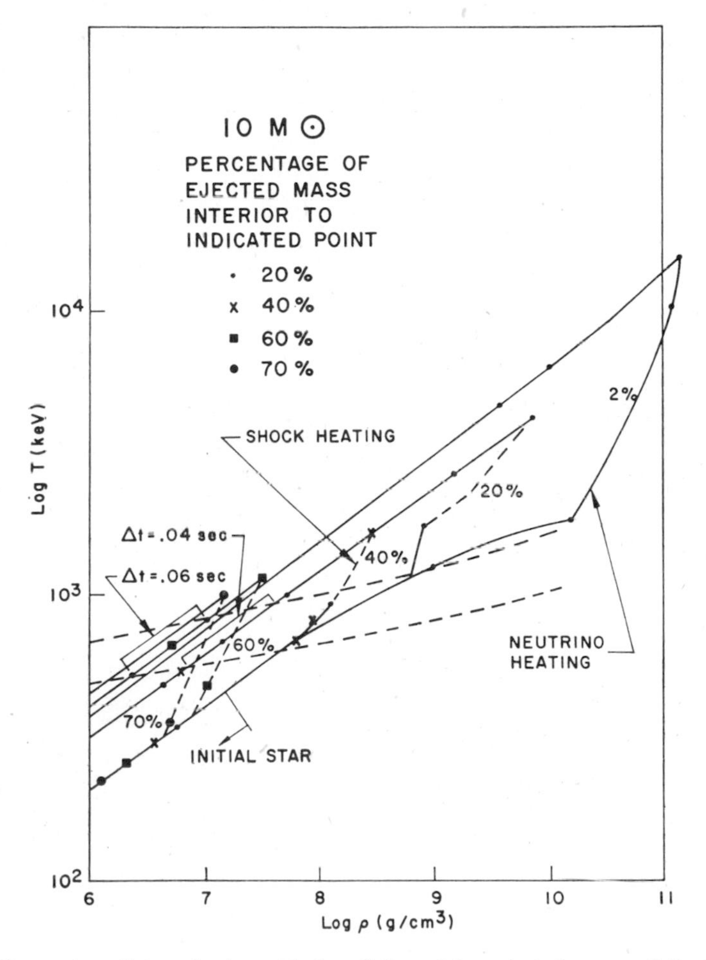

Fig. 9. Shows the adiabats in the ϱ, T plane followed by selected zones of the 10 M$_{\odot}$ problem.

analytic form for the equation of state to be expected from consideration of the processes of iron-helium conversion, helium-to-proton-neutron conversion, degeneracy and radiation pressure. The neutrino fraction of the specific heat was not included. Figure 9 shows the initial implosion, core formation, and neutrino energy transfer, and the heating and subsequent cooling of the ejected material. The ejected zones are shown cooling along adiabats of $\gamma \sim 4/3$. Figures 10 and 11 show the same thing for the 2 M$_{\odot}$ and 1.5 M$_{\odot}$ models. The time scale is of the order of 0.04 seconds.

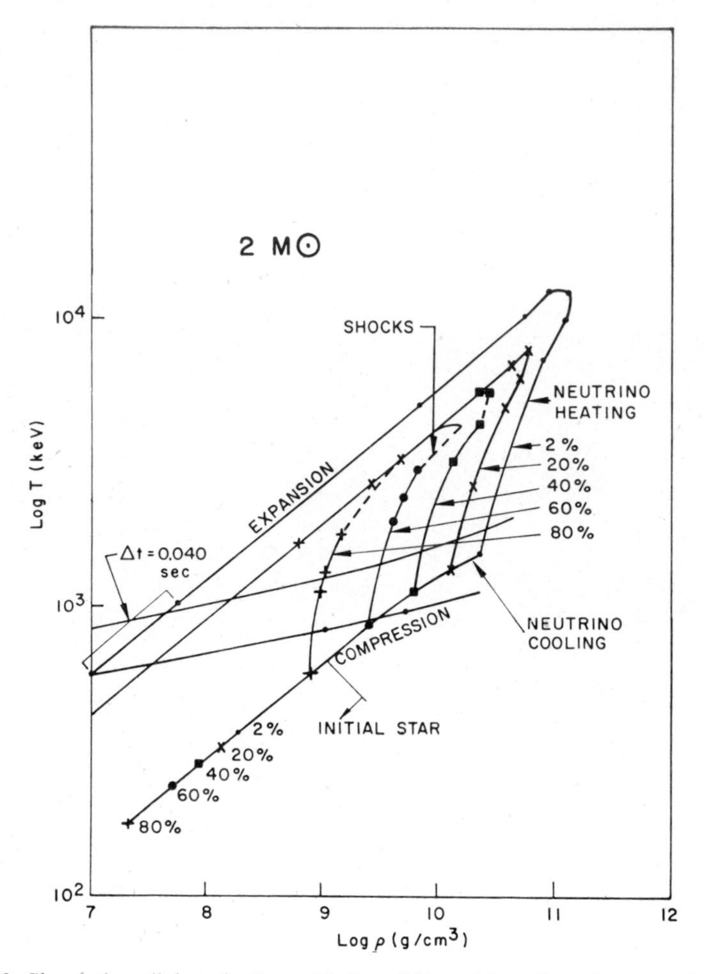

Fig. 10. Shows the adiabats in the ϱ, T plane followed by selected zones of the 2 M⊙ problem.

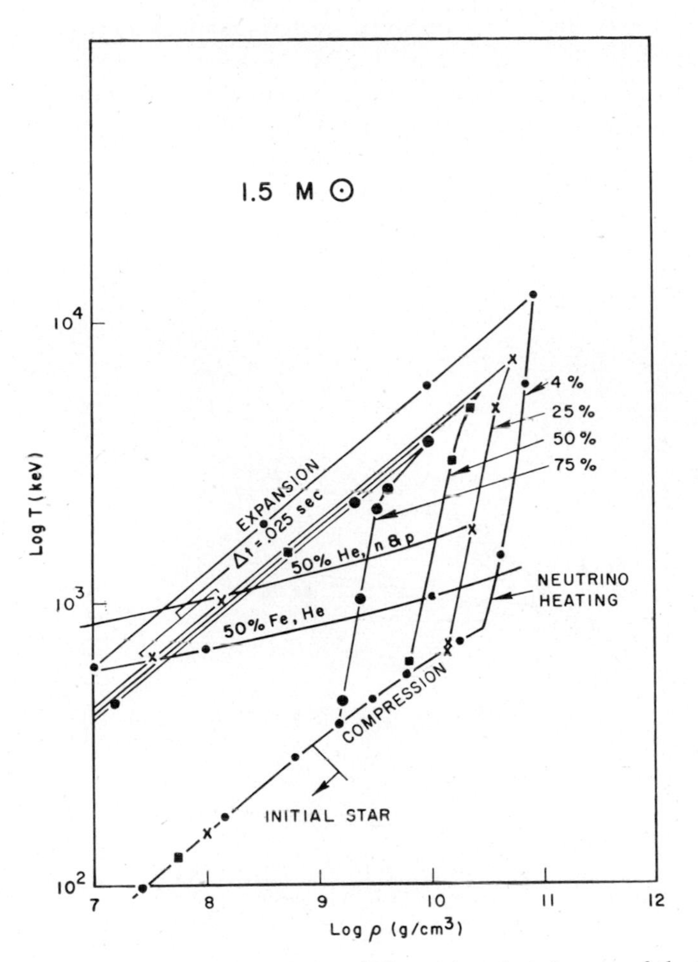

Fig. 11. Shows the adiabats in the ϱ, T plane followed by selected zones of the 1.5 M⊙ problem.

NUCLEOSYNTHESIS BEHIND A SUPERNOVA BLAST WAVE

Taking three different temperatures corresponding to ejection of material of density $\varrho \sim 10^{11}$ gm/cm^3 from 1.5 M$_\odot$ to 10 M$_\odot$ supernovae, equilibrium neutron-proton ratios were calculated. Figure 12 shows this in the $\varrho - T$ plane. The calculated neutron-proton ratios use the work of Chris

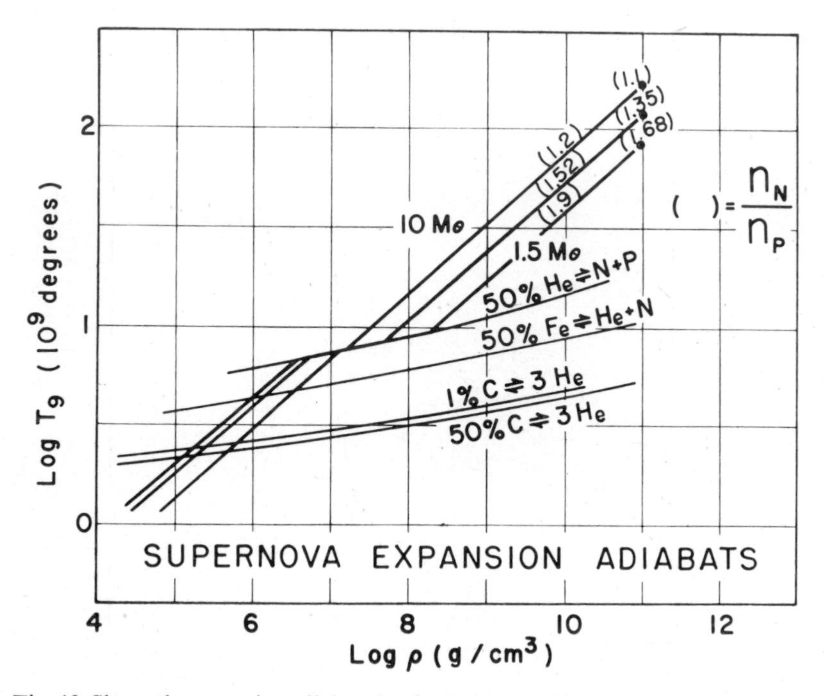

Fig. 12. Shows the expansion adiabats for the 10 M$_\odot$ problem for the fraction of matter starting at a density of 10^{11} g/cm^3. The different adiabats are for slightly different temperatures and resulting larger neutron-proton ratio differences. The lines of constant fractional He, Fe, C content are also shown along with the effect on the adiabats of the exothermic heat of the reactions.

McKee (1964) on the equation of state of high-density high-temperature matter with conservation of lepton number. Under these conditions the material is inside of the neutrino emitting surface, so that the sum of all electrons and electron-type neutrinos is conserved during the process of heating and initial expansion.

In the previous case where inverse beta-decay was considered during contraction, neutrinos or antineutrinos escaped until the matter was in equilibrium with respect to weak interactions. In the present situation the neutrinos (or antineutrinos) do not escape, but the equilibrium occurs with a finite number density of neutrinos. As the material expands, some density is reached at which the weak interactions are "frozen"; this occurs before the loss of neutrinos is sufficient for the number density of neutrinos to be negligible.

When the material is heated the neutron-proton ratio is roughly 1.7 while the number becomes 1.9 when the matter is frozen. This is for a lepton-to-baryon ratio of one-quarter, which assumes one-quarter of the leptons are lost relative to the condition of the material when it was heated.

The nucleons in the material are in the form of free protons and neutrons (the density of the material is considerably less than nuclear density). Binding begins with neutron-proton burning to form deuterium and consequently helium. This occurs at a temperature $T \sim 8 \times 10^9$ °K. When the expansion adiabats shown in Figure 12 strike the helium-forming region, the large energy release of these reactions forces the matter to follow a curve of state which maintains the helium-to-proton-neutron ratio until the exothermic energy is used up in fluid-dynamic work PdV. More simply, the exothermic reactions which form helium tend to resist changes in composition and temperature until the energy release is dissipated by doing macroscopic work. This tends to bunch the adiabats more closely as they emerge from the process of helium formation. For the adiabats considered, there will be roughly two neutrons per helium nucleus after helium formation.

Helium burning to carbon occurs at much lower temperature and density than the iron-helium-neutron phase change. This is simply because the binding per alpha particle in carbon is only about 2.5 MeV, but is 6 to 7 MeV in iron. Assuming that there are no seed nuclei, no new elements are formed in appreciable abundance until the temperature falls to that appropriate to helium burning to carbon. It should be noted that the transition regions for changing the elemental compositions are very narrow in the $\varrho - T$ plane. A slight change in temperature may change the composition drastically. If the transit time is short, a large neutron excess is expected in the region of helium-burning.

The ratio of the reaction time for the three-alpha process

$$3\alpha \rightarrow C^{12}$$

to the expansion time in this region of temperature and density if of the order of ten to one. In other words, only ten percent or less of the helium will have burned to carbon as the matter expands with the velocities involved.

In the case of the 1.5 $M_\odot$ star, the matter is coming out at a temperature $T \sim 2.5 \times 10^9$ °K and a density $\varrho \sim 10^9$ °K and an exposure time $\Delta t \sim \frac{1}{3}$ to $\frac{1}{2}$ second. The 10 $M_\odot$ star, with its higher temperature expansion adiabat, does not have such a large neutron excess. It appears that the *r*-process would not be important for such massive stars which become type II supernovae. The smaller mass, type I supernovae are characterized by large neutron excess, lower temperature adiabat at higher density, and more rapid burning to carbon. The transit time is about 0.3 seconds for the 1.5 $M_\odot$ star.

These characteristics vary as a function of the mass fraction chosen for consideration. The adiabats for these three stars are for the 5% mass fraction. For zones further out, the processes may be quite different because the adiabats will not be the same. For zones further in the conditions do not change drastically. In other words, the inner 5% mass zone gives a reasonable representation of the adiabats of the matter.

REFERENCES

Colgate, S. A. and White, R. (1966). *Astrophys. J.* **143**, 626.
McKee, C. (1964). Equilibrium Properties of Matter at Very High Density and Temperature. (To be published).

DISCUSSION

Dr. Burbidge: How uncertain is this time scale of 0.3 seconds? I am a little lost, really, as to how you arrived at it.

Dr. Colgate: It is set by taking the energy available in the helium-forming reactions and saying that it is lifting maybe five or ten times that amount of mass out of the gravitational potential well. The kinetic energy density is then

$$U^2 \sim \frac{6 \cdot 10^{18} \text{ ergs/gm}}{10}$$

which gives an expansion velocity

$$U \sim 10^9 \text{ cm/sec}$$

It is this dominant quantity of late energy which determines the expansion velocity. If 10 solar masses of matter is expanded at 10^9 cm/sec, it takes about a second to traverse the density range in which we are interested.

Dr. Reeves: Tell me if I am right. I thought that in order to explain the last peak in the abundance curve by the *r*-process, a time of the order of 1.4 seconds is required.

Dr. Seeger: That will come out in my talk later in this conference. I find that times of the order of three or four seconds are required.

Dr. Colgate: Is that limit by beta-decay?

Dr. Seeger: Yes, by all of them.

Dr. Cameron: And this starts from iron, not from carbon as you do, which would be a more stringent requirement.

Dr. Colgate: The burning up to silicon is rapid compared to the burning through carbon. I think the silicon forms seed nuclei when the neutron excess is 200.

Dr. Cameron: At 2.5 billion degrees your time scale is too short to get that far by charged particle reactions.

Dr. Colgate: The short time scale (0.3 seconds) referes to the majority of the matter. As is evident from Figure 3 a smaller, inner, fraction of the matter expands at a very much reduced velocity. It may be this small mass fraction that is envolved with the longest neutron exposure and synthesis time that we recognize as *r*-process elements.

Hydrodynamic Implosion Calculations

W. DAVID ARNETT

Various authors have suggested that type II supernovae are initiated by gravitational collapse of the central region of a star[1]. Colgate[2] and his collaborators have investigated a mechanism for supernovae by numerical integration of the equation of hydrodynamics by the Von Neuman-Richtmeyer method of pseudoviscosity. Because this technique is somewhat foreign to astrophysical literature, some problems encountered and results obtained in a preliminary investigation of the supernova implosion will be discussed.

Restrictions upon zone size

If a stellar model is divided into spherical zones of a given mass (a "Lagrangian" coordinate system), the differential equations of hydrodynamics for the model may be replaced by difference equations for the motion of the zones[4]. In particular, the momentum equation may be written as

$$\frac{\partial U}{\partial t} = -\frac{GM_r}{R^2} - 4\pi R^2 \frac{\partial P}{\partial M_r} \tag{1}$$

where

$$M_r = \int_0^r 4\pi r^2 \varrho \, dr .$$

This last quantity is not a function of time in these coordinates; P is the pressure, ϱ the density, R the radius and U the velocity of a fluid particle

[1] See Burbidge (1957), Cameron (1958), Chiu (1961) for instance.

[2] A review of this work is Colgate and White (1964).

[3] Subsequent work (to be published), through June 1965, appears in the Author's thesis, Yale University, 1966.

[4] For details of the technique, see Colgate and White (1964), Arnett (1965), and Christy (1964).

at some time t, r is the value of R at some reference time $t = 0$ and G is the gravitational constant. Transforming to difference equations, the derivatives in Equation (1) can become

$$\frac{\partial U}{\partial t} \rightarrow \frac{U(t + \Delta t) - U(t)}{\Delta t} = \frac{\Delta U}{\Delta t}$$

$$\frac{\partial P}{\partial M_r} \rightarrow \frac{P(M_r + \Delta M_r) - P(M_r)}{\Delta M_r}$$

$$= \frac{\Delta P}{\Delta M_r}.$$

If the pressure P changes rapidly with time (for example, when strong shock waves are present[5]), then neighboring zones should have nearly the same mass. Otherwise

$$\frac{\Delta P}{\Delta M_r}$$

no longer gives the properly weighted value of

$$\frac{\partial P}{\partial M_r}$$

and numerical results become inaccurate for $\frac{\Delta U}{\Delta t}$ and hence for the fluid velocity at the new time epoch $U(t + \Delta t)$. A rule of thumb for supernova calculations is that for adjacent zones i and $i + 1$,

$$\frac{M_i + 1 - M_i}{M_i} \lesssim 0.1$$

be satisfied[6].

A minimizing restriction on zone size is imposed by the Courant condition:

$$\Delta t < \frac{\Delta x}{c_s}$$

where Δx is the zone width, c_s the velocity of sound in the fluid. This condition states that if finite disturbances are present in the fluid, then the time

[5] The lack of strong shocks may explain why Christy (1964) could used neighboring zones with relatively large differences in mass.

[6] In the case of plane symmetry, this reduces to the requirement of Colgate and White (1964) of about ten zones per density scale height.

step Δt for numerical integration of the hydrodynamic equations must be less than the time needed for a sound wave to traverse any zone. Mathematically speaking the difference equations become unstable and give nonphysical results. For a given problem the computational time required goes roughly as the square of number of zones, so that the Courant condition may determine the number of zones to be used. For the supernova problem, a 100 zone model is roughly the limit for an IBM 7094, for instance; larger models require too much computing time.

These two zoning requirements severely restrict numerical analyses of supernova dynamics. For illustration, we consider some preliminary model calculations. Because of their oversimplified nature, the reader is cautioned not to draw detailed conclusions from these models.

PRELIMINARY SUPERNOVA MODELS

A number of order of magnitude estimates will be used to illustrate the content of the more detailed hydrodynamic calculation. A massive star in the late stages of evolution is likely to undergo a catastrophic collapse (see the articles by Chiu and by Colgate in this volume). This may occur, for example by endothermic capture by nuclei of electrons at high density ($\varrho \lesssim 10^{11}$ gm/cm^3) or endothermic photo-disintegration of nuclei ($T \gtrsim 5 \times 10^9$ °K, depending weakly on density). The central regions of the star contract almost in free-fall because the pressure is no longer adequate for support. The collapse is halted when the central regions reach a density greater than that of nuclear matter. The outer layers now rain down upon the central region which is no longer contracting, and because the flow is supersonic a shock wave is set up. The kinetic energy of the infalling material is roughly equal to the potential energy released, so that the fluid velocity before the shock is

$$v \approx \sqrt{\frac{2GM}{R}}.$$

Now the radius for which a singularity appears in the Schwarzschild solution of the general relativistic equations is

$$r_s = \frac{2GM}{c^2}$$

so that

$$v \sim \sqrt{\frac{r_s}{r}}\, c$$

The central region halts its collapse at a density of

$$\varrho \sim \tfrac{1}{2} \times 10^{15} \text{ gm/cm}^3$$

initially. This estimate is obtained from an independent particle model of nuclear forces (the inclusion of a nuclear "hard-core" potential term does not seem to affect this result much). The core initially formed has a mass

$$M \sim 1 \text{ M}_\odot$$

Now the actual position of the shock front (as obtained from a 100 zone hydrodynamic model) is

$$r \sim 10^{-2} r_g$$

so

$$v \sim 0.1c$$

Most of the kinetic energy is carried by the nucleons, which have an energy of ~10 MeV relative to the core material. Because of their relatively large mean-free path, neutrinos are the particles most likely to cool the core by carrying away energy. Assuming that neutrinos escape the star once they are formed, an approximate energy loss rate for the process of electron-positron annihilation

$$e^+ + e^- \rightarrow \bar{\nu}_e + \nu_e$$

is given by Chiu and Stabler (1961).

$$Q = -4.3 \times 10^{15}(T_9)^9/\varrho \text{ ergs/gm/sec} \qquad (3)$$

where T_9 is temperature in units of 10^9 °K and ϱ is the density. This rate does not depend upon the nature of nuclear forces but only upon the weak interaction. Using a noninteracting particle approximation for the nuclear component of the equation of state (the zero temperature limit is shown in Figure 1), the history of a 10 $\text{M}_\odot$ polytrope of index 3 was followed from collapse initiated by the nuclear photodisintegration mechanism. A 0.5 $\text{M}_\odot$ core formed, and was shock-heated, but the energy loss rate given by (3) was sufficient to prevent the shock from propogating outward and expelling outer layers of the star. Since this neutrino energy loss process may not be the most severe that can occur in the hot, dense core, this result suggests that no matter will be shock-ejected in this model. More matter rained down on the core until its mass[7] was ~2.5 $\text{M}_\odot$ and there was no sign of mass ejection.

[7] Special relativistic effects in the nuclear equation of state were neglected.

However, the previous neglect of neutrino interactions as they stream out of the star is unjustified. Recalling the previous estimate of the kinetic energy of the infalling nucleons relative to the core, neutrinos are expected to be formed with an average energy of the order of

$$\bar{\varepsilon} \sim 10 \text{ mev.}$$

Bahcall and Frautschi (1964) estimate the cross-section for inelastic scattering with degenerate electrons to be

$$\sigma \sim 2 \times 10^{-44} \left(\frac{E}{m_e c^2} \right)^2$$

The process is

$$\nu_e + e \rightarrow \nu_e' + e' \tag{2}$$

and

$$\bar{\nu}_e + e \rightarrow \bar{\nu}_e' + e' \tag{3}$$

For antineutrino scattering the results are multiplied by 1/3. The work of Tsuruta (1964) indicates that at the density of the core, the electron Fermi energy is

$$\varepsilon_f \sim 100 \text{ MeV}$$

or the number density of electrons is

$$N_e \sim 10^{37} \text{ cm}^{-3}$$

Since the temperature behind the shock front is

$$kT \sim \bar{\varepsilon} \sim 10 \text{ MeV}$$

then

$$\frac{\varepsilon_f}{kT} \gg 1$$

so the electrons are degenerate. The mean free path for a 10 MeV neutrino is then

$$\begin{aligned} l &= \frac{1}{N\sigma} \sim \frac{1}{10^{37} \cdot 2 \cdot 10^{-44} \cdot 4 \cdot 10^2} \\ &= \tfrac{1}{8} \cdot 10^5 \sim 10^4 \text{ cm.} \end{aligned}$$

This is an underestimate, however. Because of the limited phase space available to the electrons in the exit channels of reactions (2) and (3), most interactions occur at densities

$$\varrho \ll \varrho_{\text{core}}$$

where the average neutrino energy is above the electron Fermi energy. The width of this region is much greater than the mean free path ($\Delta R \gtrsim 10^6$).

Several approximate methods (Some similar to the method of Colgate and White, 1964) of treating the neutrino energy transfer were attempted, but the results depended too sensitively upon initial assumptions to be completely convincing. A more careful treatment, using the diffusion approximation and realistic neutrino opacities is now underway.[8]

REFERENCES

Arnett, W. D. (1965). *Ph. D. Thesis*, Yale University, unpublished.

Bahcall, J. N. (1964). *Phys. Rev.* **136**, B1164.

Bahcall, J. N. and Frautschi, S. C. (1964). *Phys. Rev.* **136**, B1547.

Burbidge, E. M., Burbidge, G. R., Fowler, W. A. and Hoyle, F. (1957). *Rev. Mod. Phys.* **29**, 547.

Cameron, A. G. W. (1958). *Ann. Rev. of Nucl. Sci.* **8**, 299.

Chiu, H. Y. and Stabler, R. C. (1961). *Phys. Rev.* **122**, 1317.

Chiu, H. Y. (1961). *Ann. of Phys.* **15**, 1; **16**, 321.

Christy, R. (1964), *Rev. Mod. Phys.* **36**, 555.

Colgate, S. A. and White, R. H. (1964). "*The Hydrodynamic Behavior of Supernovae Explosions*," UCRL-7777.

Tsuruta, S. (1964). *Ph. D. Thesis*, Columbia University, unpublished.

Note added in proof: Subsequent analyses of this problem have been published since this paper was delivered. A short list follows.

Arnett, W. D. (1966). *Can. J. Phys.* **44**, 2553.

(1967). *Can. J. Phys.* **45**, 1621.

Colgate, S. A. and White, R. (1966). *Astrophys. J.* **143**, 626.

DISCUSSION

A. G. W. Cameron: I think that one of the important questions[9] which has emerged from this conference is whether the iron peak is formed in a star before it implodes in the supernova stage. If not, the observed abundance of iron-peak elements in nature requires some alternate mechanism for their formation.

In forming the iron peak, the work of Fowler and Hoyle, and others indicates a time scale of 10^3 or 10^4 seconds is needed for the beta decays necessary to increase the number of neutrons relative to protons in the

[8] Arnett (1965).

[9] See the previous paper by A. Gilbert (Ed. note).

system so that Fe^{56} is the most abundant iron isotope. On the other hand, for a system which is rapidly imploded there are two processes which can increase the number of neutrons. One is the electron capture which accompanies the higher densities; the other is neutrino capture associated with the neutrino energy transfer forms the supernova core.

If the infalling material is rich in free protons, or proton-rich nuclei like Ni^{56}, the enhanced capture of antineutrinos relative to neutrinos will cause a rapid change to a more neutron-rich composition. When this material is ejected from the star, some iron-peak material might be able to freeze out during the expansion and subsequent cooling.

These suggestions are more in the nature of questions to be investigated rather than plausible theories, but I believe some research is needed in this direction.

Neutron Capture Cross Section Measurements

J. H. GIBBONS and R. L. MACKLIN
Oak Ridge National Laboratory

Detailed analyses of the role of slow neutron capture in the synthesis of heavy elements demand a knowledge of the neutron capture cross sections for these nuclei. In this paper, I will present the results of measurements of neutron capture cross sections performed at Oak Ridge over the past several years (Macklin and Gibbons, 1965). As a means of introduction, let us consider briefly the historical development of this subject.

The abundance measurements of Goldschmidt (1937) revealed that the general features of the observed distribution of element abundances are not correlated with their chemical properties. His conclusions were confirmed by further studies by Brown (1949) and by Suess and Urey (1956) based not only on measurements of material in the earth's crust but also on data from meteorites and the solar atmosphere.

Hughes (1946) presented a paper on neutron activation cross sections of a large number of elements for pile neutrons ($\bar{E} = 1$ MeV). From a study of these results Alpher, Bethe, and Gamow (1948) noted that there is an approximate inverse relationship between neutron capture cross sections and relative abundances. They advanced a theory which pictured the early stage of matter as a compressed neutron gas. The expansion of this neutron fluid would result in the decay of some of the neutrons into protons and electrons. Subsequent neutron captures would then build up all of the heavier elements, with some readjustment due to beta decay. While there were many problems associated with this picture, the correlation they established between the neutron capture cross sections and the element abundance distributions indicated the probable occurrence of some neutron capture process in the synthesis of the heavy elements.

The period of element formation, on this model, was only about 15 minutes or so. This was consistent with the belief, at that time, that the observed element abundance distribution was uniform throughout the galaxy. The recognition that nucleosynthesis was a continuing process in stellar interiors followed the discovery by Merrill (1952) of the presence of technetium in the atmospheres of S-type stars (red giants). As technetium has no stable isotopes, its presence in abundances sufficient to be observed suggested that element synthesis was indeed taking place in those stars. Furthermore, Burbidge *et al.* (1956) noted that there was a correspondence between the spontaneous-fission half-life of Cf^{254} and the characteristic decay time of light from supernova explosions. This indicated that rapid multiple neutron capture may occur in stellar explosions.

The conclusion, drawn from these observations, that elements are being synthesized continuously in stars, was confirmed by the discovery that there are abundance differences between certain broad classes of stars, in the sense that the ratio of the abundances of the heavy elements to hydrogen was variable and this ratio could be correlated with the age of the star. It was found that the ratio of all elements with $Z > 2$ to hydrogen was an increasing function of the time of formation of the star in the galaxy.

Burbidge *et al.* (1957) have discussed in detail the various nuclear processes which play a role in the synthesis of elements in stellar interiors. According to this picture, charged-particle reactions are mainly responsible for element production through iron, beyond which neutron capture becomes the predominant mechanism. Two quite different neutron processes are assumed necessary to synthesize the abundant isotopes of heavy elements. The "*r*" process occurs on a rapid time scale (the beta decay half-lives of the product nuclei are longer than the half-lives for destruction by neutron capture) and results in the build-up of very neutron-rich isotopes, which subsequently can undergo beta decay. This process was thought to take place in supernovae. If, however, the half-life for beta decay is shorter than the half-life for neutron destruction, then this decay can be assumed to take place before the capture of a neutron and the capture path for this process (the "*s*" process) follows the valley of beta-stability, as illustrated in Figure 1. Correspondingly, the "*r*" process path lies far on the neutron rich side of the beta stable region, While most nuclei can be formed by both of these processes, there are many which can be formed solely by one process.

Clayton, Fowler, Hull, and Zimmerman (1961) and Clayton and Fowler (1961) showed that if one chooses *s*-process nuclei and plots the product of observed abundance times the measured capture cross section near 25 keV against the mass number one obtains a remarkably smooth curve. Its detailed shape contains the "history" of the *s*-process, since it is a measure of the time integrated neutron flux to which "seed" nuclei, asssumed to be Fe^{56}, have been subjected. Such a correlation does not exist for the products of *r*-process neutron capture.

It is clear that a true test of the theory of *s*-process synthesis demands an accurate knowledge both of the neutron capture cross sections and of the relative abundances of neighboring nuclei in the slow capture path. As the relative abundances can be distorted by indeterminable amounts of physical and chemical fractionation, it is preferable to study the correlations of cross section times abundance for various *s*-process isotopes of a single element. In this case, the isotopic abundances are quite accurately known, and it is necessary only to determine the neutron capture cross sections to sufficient accuracy. I would like now to present the results we have obtained for such correlations from our neutron capture cross section data.

Figure 2 shows a portion of the slow neutron capture path illustrated in Figure 1. We have plotted atomic number versus mass number so that the *r*-process β-decay is vertical. The dashed line traces the *s*-process path in this region. For the case of tin, we see that the isotopes with $A < 116$ have not resulted from *s*-process synthesis. The heavy isotopes, $A \geqq 122$, should result from *r*-process synthesis only, due to the short beta decay half life of Sn^{121}. Sn^{116} lies directly on the *s*-process capture path, and is shielded from any contribution from the *r*-process by the beta stable Cd^{116}. The contribu-

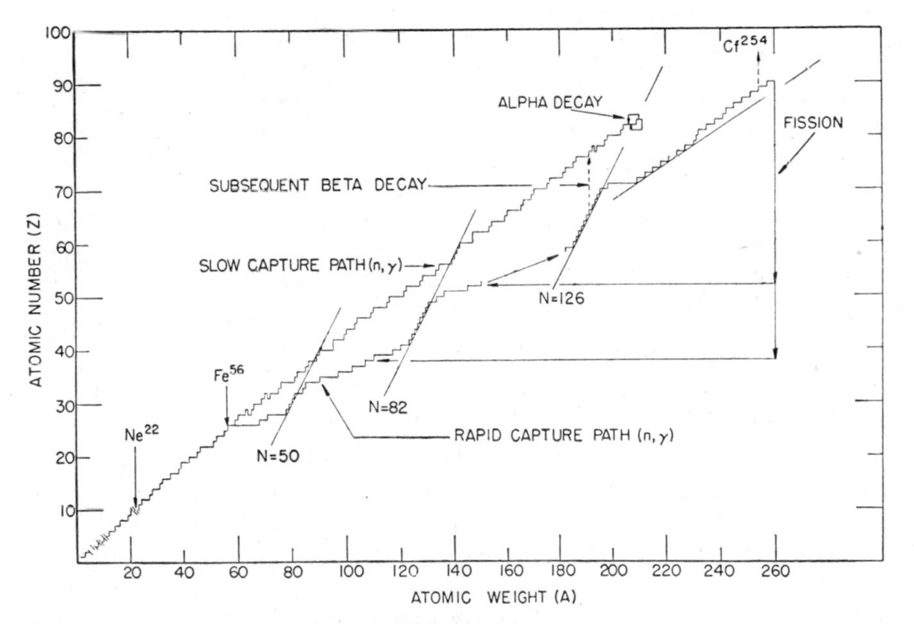

Fig. 1. Schematic diagram of the neutron capture paths for *s*-process and *r*-process synthesis.

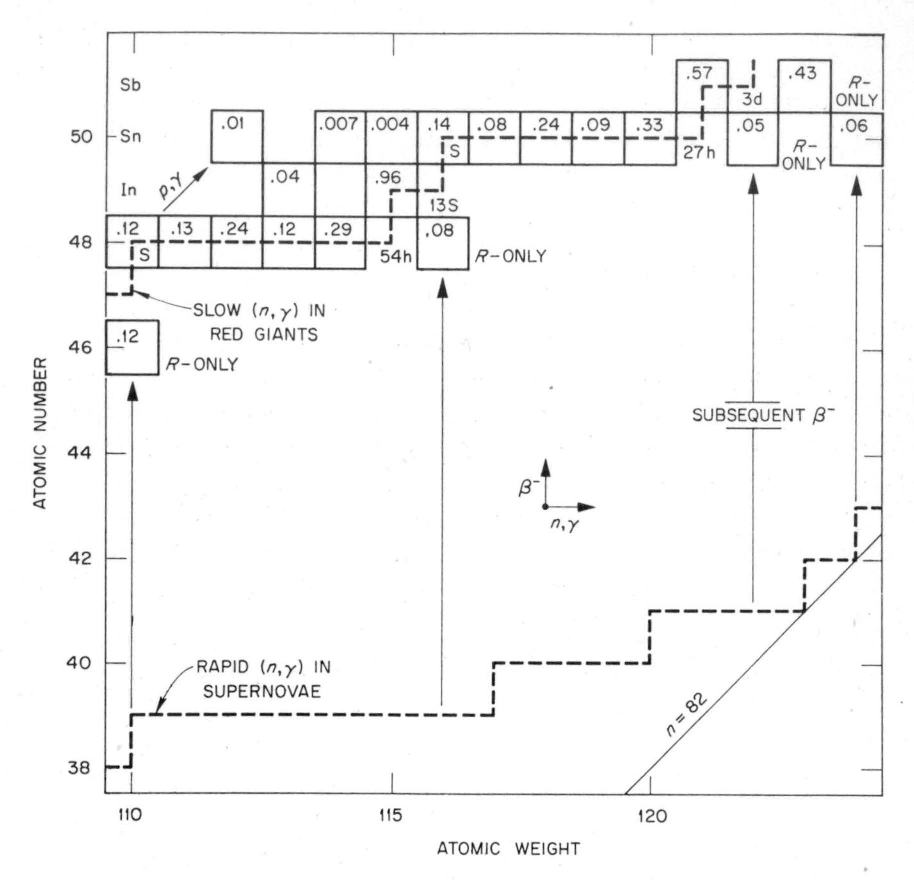

Fig. 2. The slow neutron capture path (indicated by the dashed line) is shown in the vicinity of tin. The atomic number is plotted as a function of mass number so that the *r*-process decay is vertical.

tion of *r*-process synthesis to the abundances of Sn^{118} and Sn^{120} is apparently rather small, while for Sn^{117} and Sn^{119} it could have been significant.

The problem, therefore, was to measure the capture cross sections for these isotopes at an energy corresponding to the conditions in the interiors of red giant stars. Under these conditions the relative velocity v between neutron and target is determined by the Maxwell-Boltzman distribution, and the reaction rates are weighted averages of the product of the velocity, v, the capture cross section, σ, the abundance, N, for each target isotope, and the local free neutron density (Clayton et al., 1961). The weighting function is

$$\varphi(v)\,dv = \frac{4}{\sqrt{\pi}}\left(\frac{v}{v_T}\right)^2 \exp - \left(\frac{v}{v_T}\right)^2 \frac{dv}{v_T}$$

where $v_T = (2kT/m)^{1/2}$ and m is the reduced neutron mass. We have calculated the average

$$\langle \sigma v \rangle = \int_0^\infty \sigma v \varphi(v)\,dv$$

from our knowledge of the energy dependence of σ (Macklin and Gibbons, 1965). The results are expressed in cross section units (millibarns) as

$$\langle \sigma \rangle = \langle \sigma v \rangle / v_T$$

for convenience.

The resulting correlations for the isotopes of tin (Macklin et al., 1962) are presented in Table I. The cross sections quoted here are calculated for $kT = 30$ keV. If the estimated r-process contributions are taken into account (Fowler, 1962) we find the product $N_s \langle \sigma_c \rangle$ to be roughly constant, decreasing by some 30% over the range $116 \leqq A \leqq 120$. The smooth variation of this product is consistent with the formation of these isotopes by s-process synthesis. However, the uncertainties associated with the magnitude of the r-process contributions render this a less than satisfactory proof of the existence of the slow neutron capture process.

Table I

Tin Isotopes at $kT = 30$ keV

A	Class	$\langle\sigma_c\rangle$mb	N (atom %)	N_r	N_s	$N_s\langle\sigma_c\rangle$
116	*s-o*	104 ± 21	14.2	—	14.2	14.8 ± 3
117	*sr*	418 ± 88	7.6	(4.0)*	3.6	15.0 ± 3
118	*sr*	65 ± 13	24.0	(4.5)*	19.5	12.7 ± 2.6
119	*sr*	257 ± 54	8.6	(4.0)*	4.6	11.8 ± 2.5
120	*sr*	41 ± 8	33.0	(4.5)*	28.5	11.7 ± 2.3
122	*r-o*	—	4.7	4.7	—	—
124	*r-o*	—	5.9	5.9	—	—

* W. A. Fowler, private communication.
Errors do not include uncertainties arising from r-process estimates.

There is another case which we felt should yield more conclusive results. The s-process capture path in the samarium region is illustrated in Figure 3.

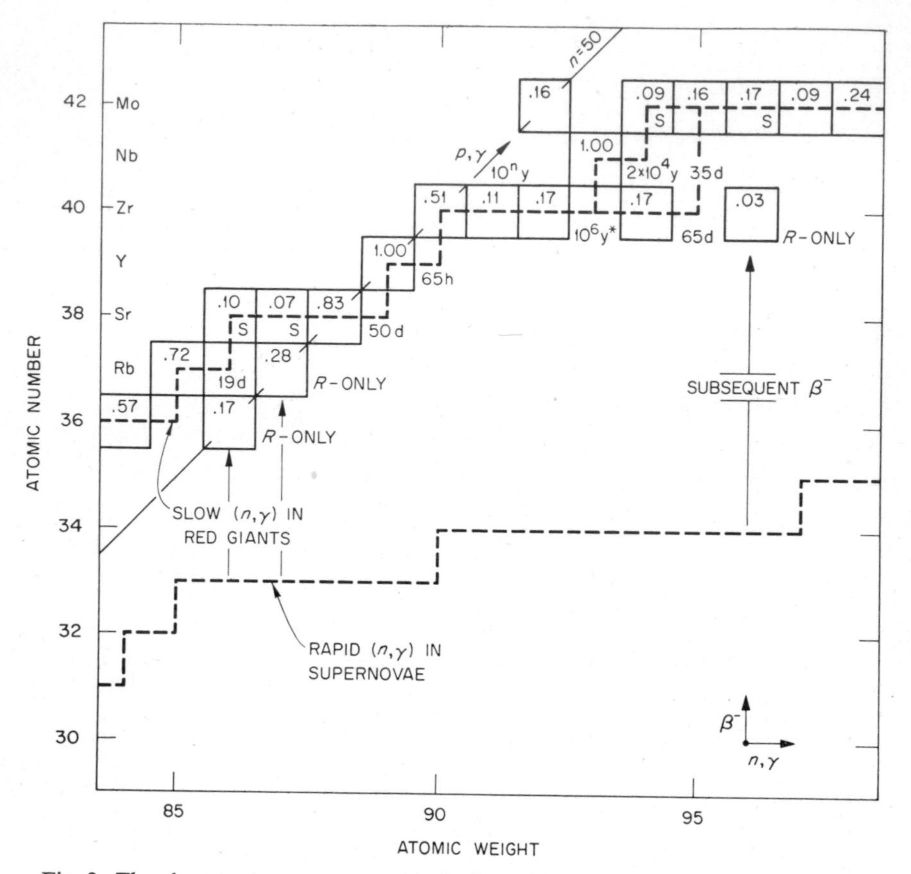

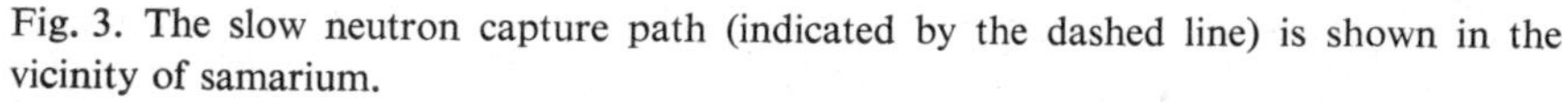
Fig. 3. The slow neutron capture path (indicated by the dashed line) is shown in the vicinity of samarium.

For samarium, one has the advantage that there are two isotopes which can be produced *only* by the *s*-process (Sm^{148} and Sm^{150}), as they are shielded from *r*-process contributions by Nd^{148} and Nd^{150}. Furthermore, in this region one would not expect the product $N_s\langle\sigma_c\rangle$ to vary rapidly with mass number. Thus, having accurately determined the neutron capture cross sections for these two isotopes, the equality of the products $N_s\langle\sigma_c\rangle$ should constitute conclusive proof of the existence of *s*-process synthesis. The results for the samarium isotopes are presented in Table II. The products $N_s\langle\sigma_c\rangle$ for Sm^{148} and Sm^{150} are seen to be equal, within the limits of uncertainty of these measurements (Macklin et al, 1963a).

We have been able, therefore, to confirm the *s*-process hypothesis by the accurate determination of the neutron capture cross sections and their

Table II

Samarium Isotopes at $kT = 30$ keV

A	Class	$\langle\sigma_c\rangle$	N (atom%)	N_r	N_s	$N_s\langle\sigma_c\rangle$
144	$p(m)$	150 ± 70	2.87	—	—	—
147	rs	1170 ± 190	14.9	12.5 ± 0.4*	2.4 ± 0.4*	(2810)*
148	s-o	257 ± 50	11.24	—	11.24	2900 ± 560
149	rs	1620 ± 280	13.85	12.1 ± 0.3*	1.7 ± 0.3*	(2810)*
150	s-o	370 ± 72	7.36	—	7.36	2720 ± 530
152	rs	410 ± 70	26.90	20.0 ± 1*	6.9 ± 1*	(2810)*
154	ro	325 ± 60	22.84	22.8	—	—

* Inferred from the ^{148}Sm and ^{150}Sm results.

correlation with isotopic abundance. Further cross section data are required however, to determine the s-process contribution to nuclei which can be formed by r-processing. This would allow an indirect measurement of the r-process contributions as a function of mass number. For example, in the case of samarium the results for Sm^{148} and Sm^{150} enable us to obtain fairly reliable estimates of the r-process for Sm^{147}, Sm^{149}, and Sm^{152}. In addition to the r-process mapping, the *details* of the s-process function (cross sections times abundance) versus atomic weight, which apparently has structure, await a better determination.

These accurate determinations of the s-process components can yield further nuclear clues to the history of our solar system abundances. For example Sm^{149} has an enormous thermal neutron capture cross section (41,500 barns) while that for Sm^{147} is only $\sim$90 barns. Assuming the s-process picture to be valid, then samarium must not have been exposed to a significant flux of thermal neutrons during the early history of the solar system (Fowler et al., 1962), or we would note an enhancement of Sm^{150} with respect to Sm^{148} (or a depletion of Sm^{149} with respect to Sm^{147}).

We have also obtained data on isotopes of strontium (Figure 4) and zirconium which provide further tests of the neutron capture s-process. Sr^{86} lies on the s-process capture path, and is shielded by Kr^{86} from any r-process production. Similarly, Sr^{87} is produced only by the s-process as it is shielded by Rb^{87}. The small cosmoradiogenic contribution from radioactive Rb^{87} can be quite reasonably taken into account. Results of this measurement are summarized in Table III. It is clear that, even with the rather large experimental errors, the $N_s\langle\sigma\rangle$ product is rapidly decreasing in

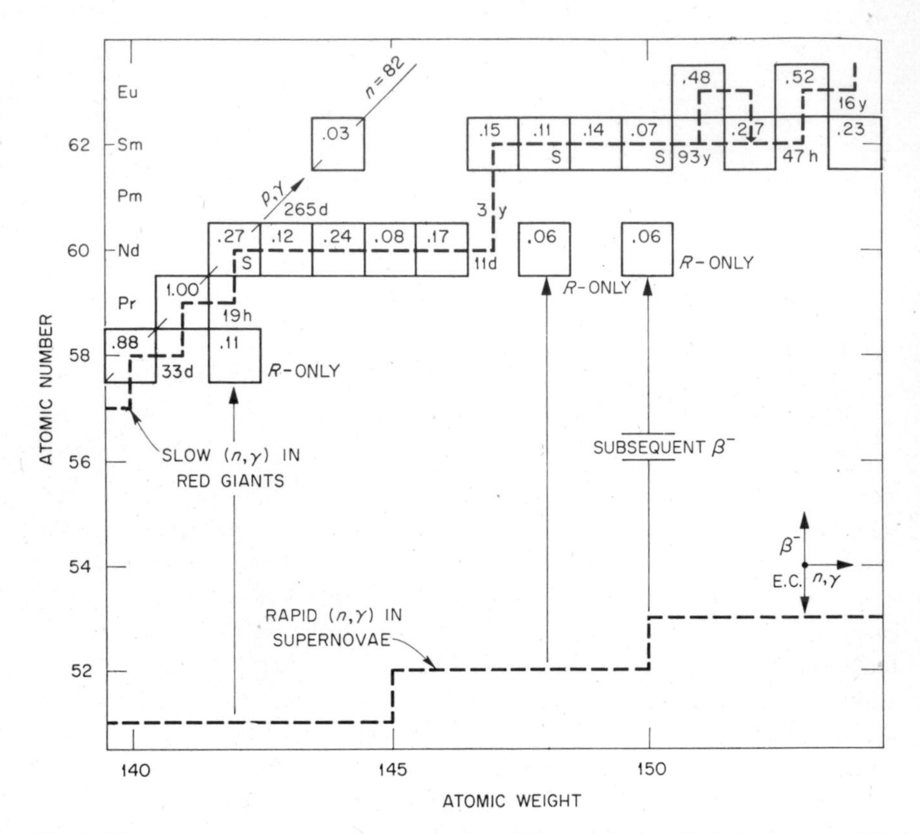

Fig. 4. The slow neutron capture path (indicated by the dashed line) is shown in the vicinity of zirconium.

Table III

Strontium Isotopes at $kT = 30$ keV

A	Class	$\langle\sigma_c\rangle mb$	N(atom %)	N_r^*	N_s	$\langle N_s\sigma_c\rangle$
86	*s-o*	75 ± 15	9.86	—	9.9	740 ± 150
87	*sr*	108 ± 20	7.02	(0.7)	6.3	680 ± 120
88	*sr*	6.9 ± 1.7	82.56	(7.8)	71.8	500 ± 130

* Seeger, P. A., Fowler, W. A., and Clayton, D. D., Astrophys. J. Supplement No. 97, Vol. XI (1965) pp. 121–166.

^{87}Sr is *s*-only but has a cosmoradiogenic decay contribution from the *r*-only ^{87}Rb.

this region. The cross section for Sr^{88} is quite small, due to the closed neutron shell, and the result is subject to large experimental errors, as noted.

Although the zirconium isotopes are not shielded from the *r*-process capture production, the contributions from the *r*-process as inferred from the isotope produced only in this manner (Zr^{96}) should be small. We felt that this would provide another test of the validity of the *s*-process. However, as in the case of the strontium isotopes and yttrium, we ran into some difficulties in obtaining these cross sections.

Strontium, yttrium, and zirconium are characterized by a magic or nearly magic number of neutrons and protons ($Z = 40$ and $N = 50$). For such nuclei, the level density is markedly decreased relative to its neighbors. In contrast, for samarium the levels are very closely spaced and the widths are broadened by the greater *s*-wave strength. Therefore, for elements in the vicinity of closed shells the properties of *individual* resonances will be important to the capture cross section. These effects are accentuated by the use of thick samples, required in order to obtain good counting statistics because of the small cross sections encountered. Thus the closure of nuclear shells (producing marked structure in the $N_s\langle\sigma_c\rangle$ *vs* A curve) has also played havoc in our ability to make accurate cross section measurements. There are satisfactory correction factors for such effects as resonance self-shielding and neutron scattering in the case of nuclei with high level density (in our case, up to about one keV). Likewise one can make reasonably accurate corrections in the case of isolated, resolved resonances. But intermediate cases are much more difficult and at present nuclei typified by Sr^{88}, Y^{89}, and Zr^{90} pose a severe problem when accuracies of better than $\pm 20\%$ are sought.

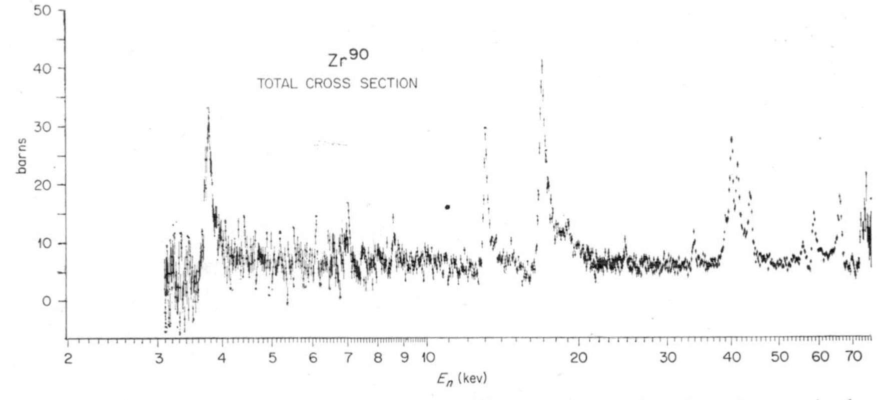

Fig. 5. The total cross section for neutrons on Zr^{90} is plotted as a function of energy in the range 2–70 keV.

The total cross section as a function of energy for Zr^{90} is shown in Figure 5. Strong individual resonances near 30 keV are apparent from this figure.

The capture cross sections for the zirconium isotopes are presented in Table IV. The associated uncertainties, due to the effects we have discussed, are as high as nearly 30%. These isotopes are believed to be produced primarily by *s*-process synthesis (Seeger, 1965). Assuming the N_r contribution to be constant, except for a small odd-even effect, we were able to obtain the products $N_s\langle\sigma_c\rangle$ for the light zirconium isotopes. The results reveal that $N_s\langle\sigma_c\rangle$ is constant (to about 10%) for the lightest three isotopes of zirconium, while it is significantly decreased for Zr^{94}. This may well indicate

Table IV

Zirconium Isotopes at $kT = 30$ keV

A	Class	$\langle\sigma_c\rangle$, *mb*	*N* (atom %)	N_r^*	N_s	$\langle N_s\sigma_c\rangle$
90	*s*(*m*)	11 ± 3	51.46	(3.0)	48.5	535 ± 160
91	*sr*	59 ± 10	11.23	(2.5)	8.7	515 ± 100
92	*sr*	34 ± 6	17.11	(3.0)	14.1	480 ± 90
94	*sr*	21 ± 4	17.4	(2.8)	14.6	310 ± 60
96	*r*-*o*(?)	41 ± 12	2.8	$\leqq 2.8$	—	—

* estimated from systematics.
errors do not include uncertainties arising from *r*-process estimates.

some significant branching at Zr^{93}. The beta decay lifetime for Zr^{93} is $\lesssim 10^6$ years. Assuming a neutron capture lifetime $\sim 10^4$ years, Zr^{93} should be a stable nucleus with respect to this process, and the product $N_s\langle\sigma_c\rangle$ product decreases smoothly from about 600 for Zr^{90} to about 300 for Zr^{96}. Better data are obviously needed in this case.

The results we have obtained for $N_s\langle\sigma_c\rangle$ for the isotopes of strontium, zirconium, tin and samarium are summarized in part in Figure 6. $N_s\langle\sigma_c\rangle$ for the one stable isotope of yttrium, Y^{89}, is also included. A rather smooth decrease with mass number is observed. With the important exception of yttrium the point scatter is small, though the cross sections involved vary by more than a factor of 10 over this range. We feel that the low value for $N_s\langle\sigma_c\rangle$ in yttrium may well be due to an inaccurate assignment of abundance, or a poor cross section measurement, or a combination of both. While the results for samarium provide the best quantitative confirmation of the *s*-process, the zirconium and strontium isotopic comparison results, together with the smooth variation which $N_s\langle\sigma_c\rangle$ evident in this figure, support this conclusion.

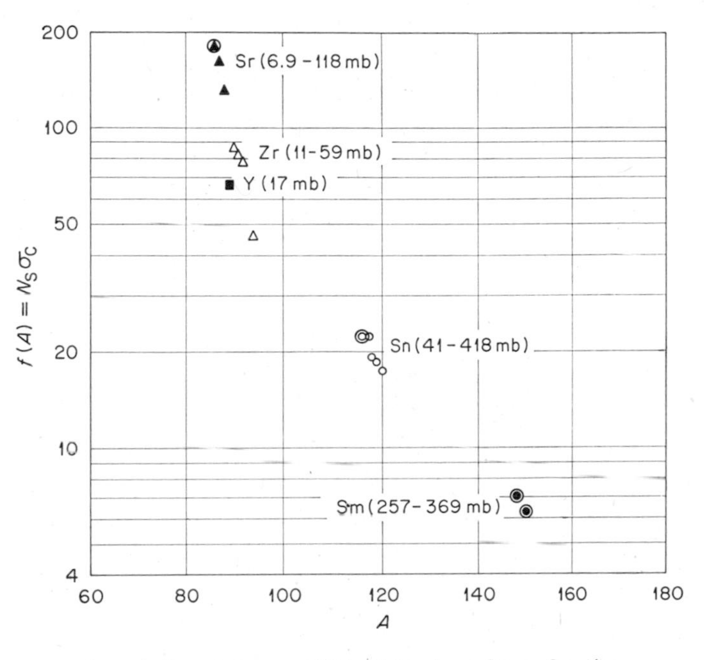

Correlation Between 30 keV Capture Cross Section
and Abundance for Certain Nuclides.

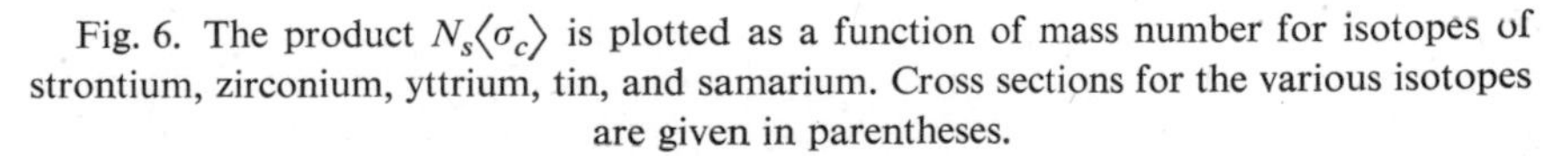

Fig. 6. The product $N_s\langle\sigma_c\rangle$ is plotted as a function of mass number for isotopes of strontium, zirconium, yttrium, tin, and samarium. Cross sections for the various isotopes are given in parentheses.

While we intend to pursue this type of analysis further, there are not very many elements that have more than one isotope produced by the *s*-process. Tellurium has three such isotopes (122, 123, and 124), but these are quite rare (2.46, 0.87 and 4.61 per cent, respectively) and it will take some time for the isotopes division at Oak Ridge to prepare the samples.*

We have continued our study of cross section systematics, with the hope that we might be able to predict, within reasonable uncertainties, the capture cross sections of nuclei for which the preparation of samples is too difficult to be feasible. Figure 7 shows results obtained for the average neutron capture cross sections at 65 keV as a function of proton number Z. The odd-even effect has been reduced by multiplying the cross sections of even-even (proton-neutron) target nuclei by a factor 2.2. This factor is empirically

Note added in proof:* Since this paper was written the tellurium cross sections have been measured and the $N\sigma$ products agree with *s*-process predictions. See Macklin and Gibbons, *Phys. Rev.* **159, 1007 (1967) and *Astrophysical Journal*, September 1967.

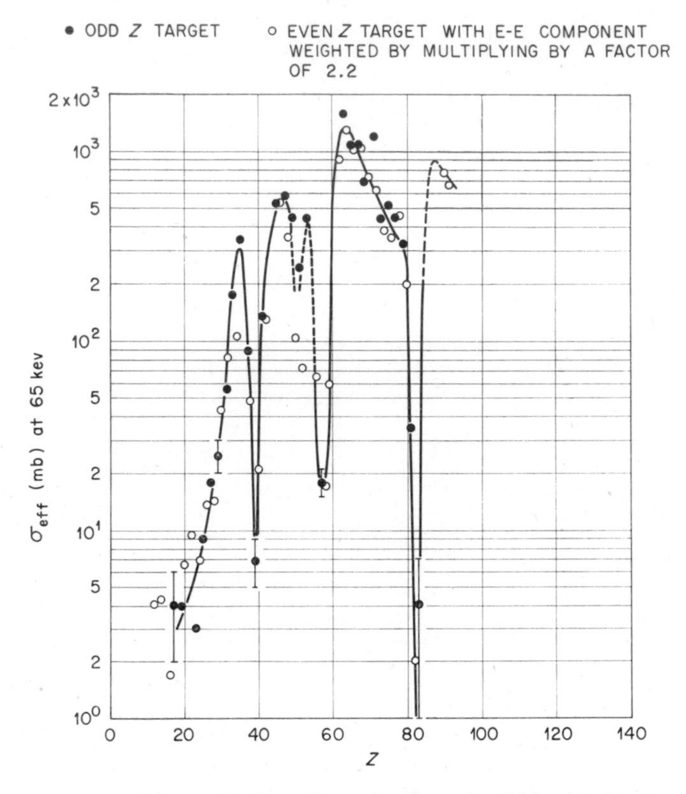

Fig. 7. Average neutron capture cross sections calculated for $kT = 65$ keV are plotted as a function of mass number. The odd-even effect has been empirically reduced by multiplying the cross section for even-even nuclei by a factor 2.2.

found to be largely independent of mass number (Gibbons et al., 1961; Macklin et al., 1963b). While the trends are somewhat uncertain in the vicinity of the closed shell at $Z = 50$ and $N = 82$, very consistent trends result elsewhere. However, it is obvious that many of the nuclides of greatest interest lie on the most rapidly changing part of the curve. For instance, the cross sections for the three strontium isotopes vary by a factor of up to 15. These trends are essentially independent of energy in this energy range ($kT \gtrsim 30$ keV) except for the lighter and doubly magic nuclei.

An extensive compilation of Maxwellian averaged neutron capture cross sections as a function of kT, for the various elements for which experimental evidence is available, has been presented in the literature (Macklin and

Gibbons, 1965). Some examples of the variations of these cross sections ($\langle\sigma v\rangle/v_T$) with kT are illustrated in Figure 8. The cross section curve for S^{32} corresponds to the contribution from a single resonance at $\sim$30 keV. For the other cases, there are several resonances in the energy range of interest. A generally smooth decrease in cross section with temperature is evident for these cases.

Currently, we are concerned with the determination of the neutron capture cross sections for the isotopes of lead and generally with the development of techniques for measuring smaller cross sections more accurately.

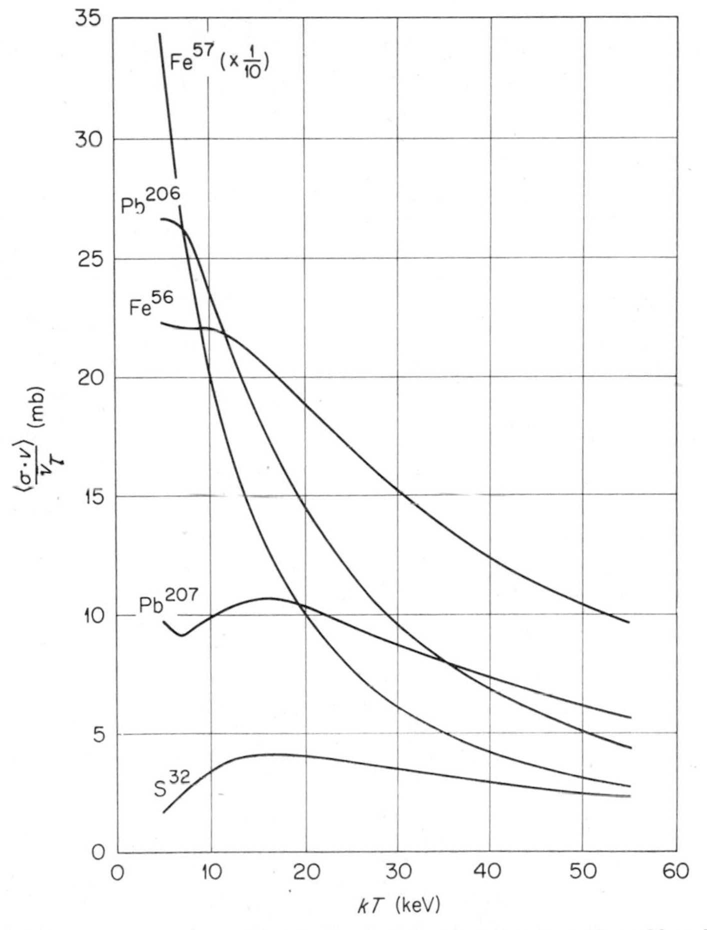

Fig. 8. The average neutron capture cross sections calculated for S^{32}, Fe^{56}, Fe^{57}, Pb^{206} and Pb^{207} are plotted as a function of kT.

These results are important both for problems of cosmochronology and for a consideration of the termination of the *s*-process. The neutron capture cross section for Pb^{207} is shown as a function of energy in Figure 9. While there was known (from transmission measurements) to be a single *s*-wave resonance near 40 keV, the capture cross section curve reveals a somewhat more complex situation.

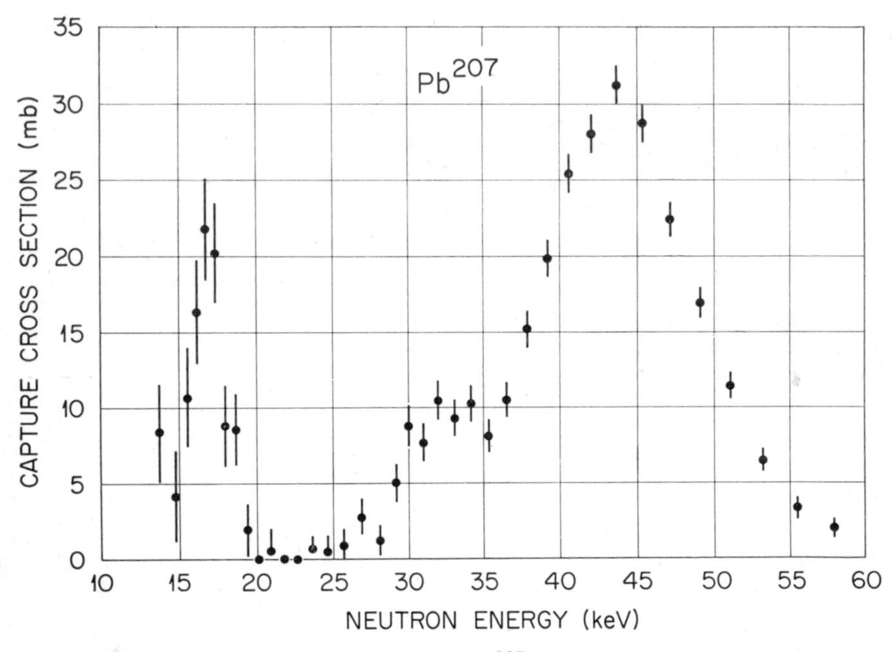

Fig. 9. The neutron capture cross section for Pb^{207} is shown as a function of energy in the range 10–60 keV.

In order to further investigate this behavior, we performed a careful measurement of the total neutron cross section for Pb^{207}. This result is illustrated in Figure 10. We note here the presence of p-wave resonances. Although these states are almost unobservable in the total or scattering cross sections, they make a very significant contribution to the capture cross section, as confirmed by the more pronounced features evident in Figure 9. This effect is similar to that observed for (α, γ) reactions on medium weight nuclei, where resonance states which have negligible reduced widths for scattering can be very important in the capture processes.

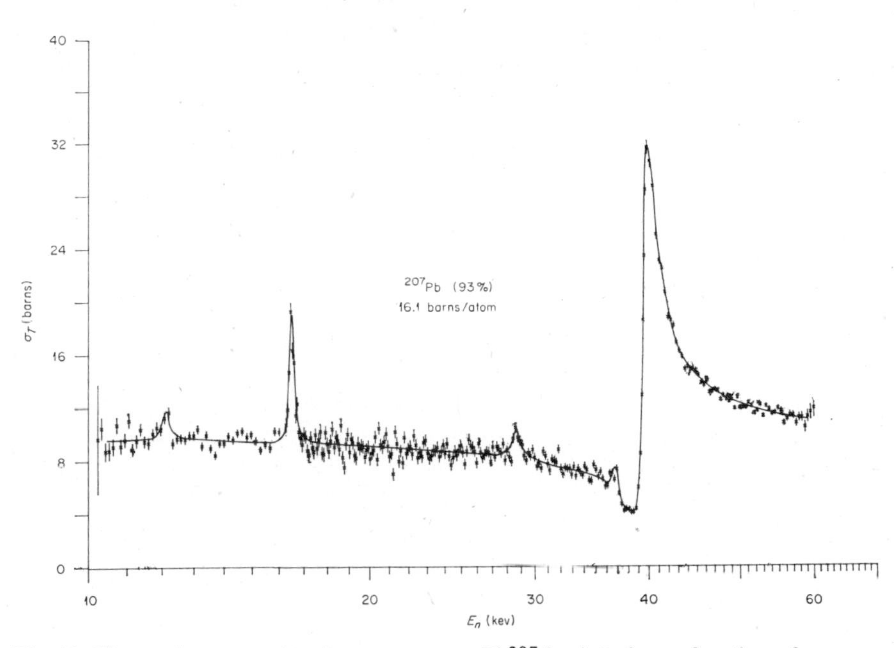

Fig. 10. The total cross section for neutrons on Pb^{207} is plotted as a function of energy in the range 10–60 keV.

The total cross section for neutrons on Pb^{208} as a function of energy is shown in Figure 11. In this energy region, what was thought to be a single resonance near 80 keV was found to be two resonances. We have not yet had a satisfactory sample for a capture measurement but we guess that these two are the only resonances below 100 keV.

A good example of the difficulties one might encounter in trying to guess the capture cross section from a knowledge of the total cross section is provided by Fe^{57}. The total cross section for neutrons on Fe^{57} is shown in Figure 12. There are no resonances apparent in these high resolution data from 10 keV to $\sim$ 30 keV. In contrast, the capture cross section for Fe^{57}, shown in Figure 13, reveals the presence of several very strong resonances in the range 10–15 keV, as well as an appreciable cross section due to unresolved resonances between 25 and 50 keV.

If we are to obtain cross section results to accuracies better than 20–25% for nuclei with spacings of a few keV, it will be necessary to improve the capture time resolution. A recent total cross section measurement for Cu^{65} is shown in Figure 14. The individual resonances are quite apparent in the vicinity of 30 keV. In this region, the resonance widths are of the order of

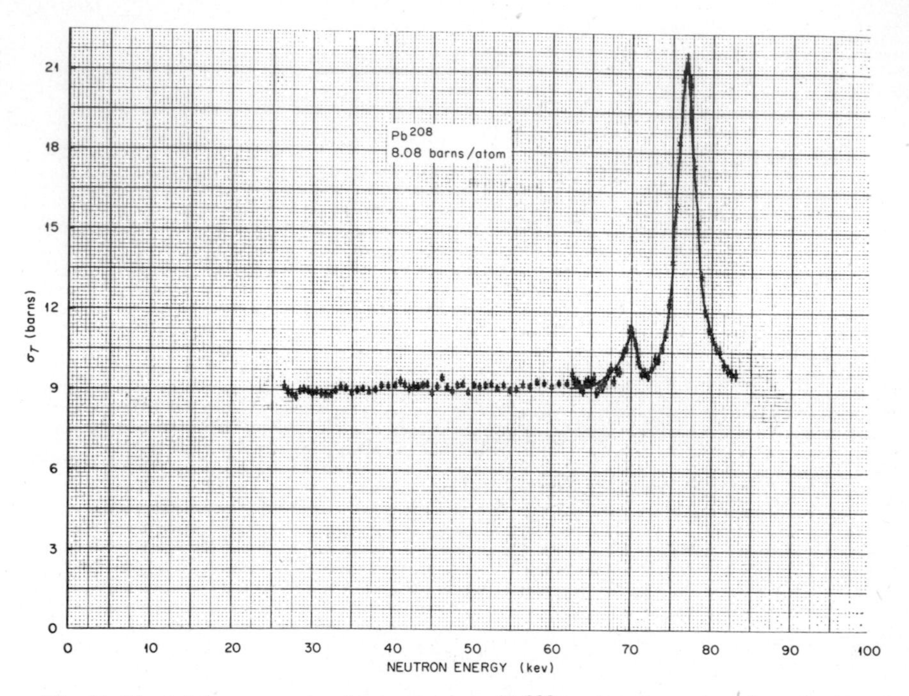

Fig. 11. The total cross section for neutrons on Pb^{208} is plotted as a function of energy in the range 25–85 keV.

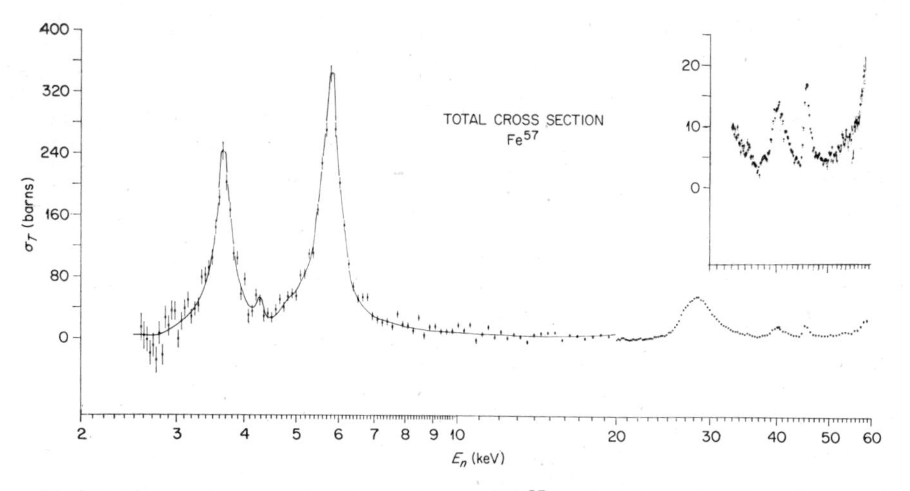

Fig. 12. The total cross section for neutrons on Fe^{57} is plotted as a function of energy in the range 2–60 keV.

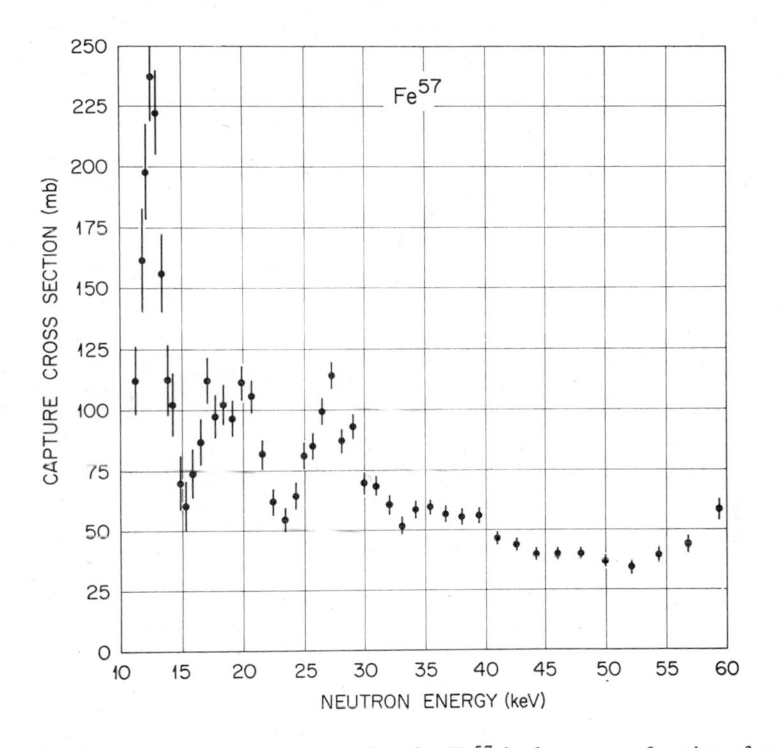

Fig. 13. The neutron capture cross section for Fe^{57} is shown as a function of energy in the range 10–60 keV.

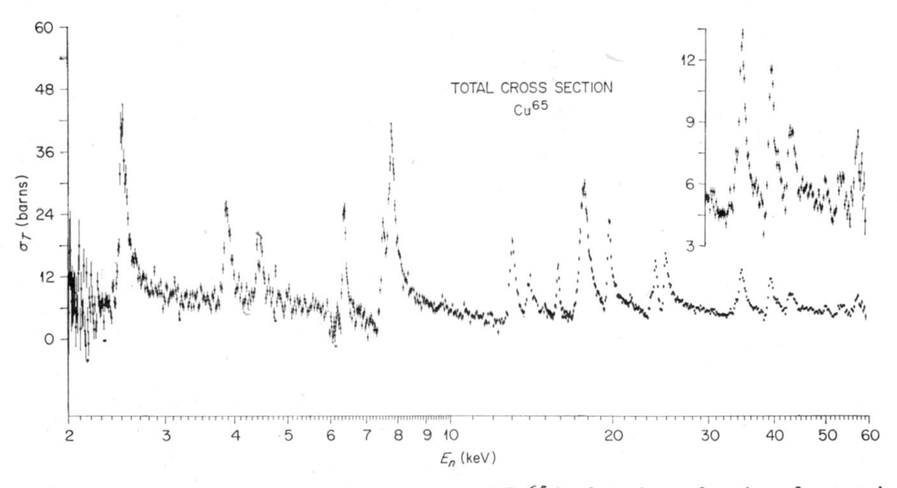

Fig. 14. The total cross section for neutrons on Cu^{65} is plotted as a function of energy in the range 2–60 keV.

hundreds of eV, while the level spacing is a few keV. We need energy resolution in the capture measurements more closely approximating that obtained in our total cross sections. Thus we are currently working on a "total gamma energy" detector which has five times the efficiency of our present detector (Macklin et al., 1963c). This, alone, should result in a resolution improvement of a factor of two. In addition we are striving for considerably more intense neutron bursts by improving the pulsed proton accelerator characteristics.

I would like, in conclusion, to mention briefly some of the difficulties associated with the preparation of samples for future neutron capture cross section determinations. Tellurium is an interesting element to investigate, as I have mentioned previously, as it has three isotopes produced by *s*-process synthesis. The problem here is that the tellurium *s*-process isotopes are rare, and the preparation of large samples is expensive and time consuming. Osmium, another important element with regard to both *s*-process synthesis and chronology, is highly toxic. As containment facilities are not available in the electromagnetic separation units at Oak Ridge, the osmium separation must be handled "after hours" with special precautions. Neutron capture cross sections for krypton and xenon are important in considerations of the variation of the observed abundances of their isotopes in the earth, sun, and meteorites. Thermal diffusion columns must be constructed in order to separate significant quantities of these rare gases. It should be clear, therefore, that future studies of *s*-process neutron capture cross sections will involve an increased amount of isotope separations effort as well as continued improvements in keV range neutron production and radiative capture detection methods.

The rewards, in increased understanding of the *s* and *r*-processes, seem well worth the effort.

REFERENCES

Alpher, R. A., Bethe, H. and Gamow, G. (1948). *Phys. Rev.* **73**, 803.

Brown, H. (1949). *Rev. Mod. Phys.* **21**, 625.

Burbidge, E. M., Burbidge, G. R., Fowler, W. A. and Hoyle, F. (1957). *Rev. Mod. Phys.* **29**, 547.

Burbidge, G. R., Hoyle, F., Burbidge, E. M., Christy, R. F. and Fowler, W. A. (1956). *Phys. Rev.* **103**, 1145.

Clayton, D. D. and Fowler, W. A. (1961). *Ann. Phys.* **16**, 51.

Clayton, D. D., Fowler, W. A., Hull, T. E. and Zimmermann, B. A. (1961). *Ann. Phys.* **12**, 331.

Fowler, W. A. (1962). Private communication.
Fowler, W. A., Greenstein, J. L. and Hoyle, F. (1962). *Geophys. J. Roy. Astron. Soc.* **6,** 148.
Gibbons, J. H., Macklin, R. L., Miller, P. D. and Neiler, J. H. (1961). *Phys. Rev.* **122,** 182.
Goldschmidt, V. M. (1937). *Skrifter Norske Videnshaps-Acad. Oslo I: Mat. Naturv.* Kl. No. 4.
Hughes, D. J. (1946), *Phys. Rev.* **70**, 106A.
Macklin, R. L., Inada, T. and Gibbons, J. H. (1962). *Nature* **194**, 1272.
Macklin, R. L., Gibbons, J. H. and Inada, T. (1963a). *Nature* **197**, 369.
Macklin, R. L., Gibbons, J. H. and Inada, T. (1963b). *Phys. Rev.* **129**, 2695.
Macklin, R. L., Gibbons, J. H. and Inada, T. (1963c). *Nucl. Phys.* **43**, 353.
Macklin, R. L. and Gibbons, J. H. (1965). *Rev. Mod. Phys.* **37**, 166.
Merrill, P. W. (1952). *Science* **115**, 484.
Seeger, P. A., Fowler, W. A. and Clayton, D. D. (1965). *Astrophys. J. Suppl.* **97**, 121.
Suess, H. E. and Urey, H. C. (1956). *Rev. Mod. Phys.* **28, 53.**

Neutron Cross Section Measurements with Nuclear Devices

P. A. SEEGER

I would like to discuss the possibility of making neutron cross section measurements using a nuclear explosion as a source of neutrons. The chief advantage to be gained from such experiments is the high flux available. A laboratory accelerator can produce on the order of 10^{21} neutrons per year, typically in bursts of 0.1 μsec duration and with a broad fission-like energy spectrum. In contrast, a modest nuclear detonation, with a yield comparable to one kiloton of TNT, produces approximately 10^{23} neutrons in a burst of duration <0.1 μsec and with a similar energy distribution. Thus this method is most useful for measuring short-lived or highly radioactive targets. With regard to nucleosynthesis, one might consider making targets of very neutron-rich fission products to measure (n, γ) cross sections involved in rapid-capture processes.

This report will describe an experiment performed at the Nevada Test Site of the Atomic Energy Commission, on December 16, 1964. The device was exploded underground. A vertical vacuum pipe 187 meters in length provided a direct line of sight to the bomb. The 12-in. diameter vacuum column contained many antiscattering baffles, and 10 ft below ground there was a steel collimator 4 ft in length, from which there emerged a 1-cm^2 beam of neutrons. The space above the collimator contained lead shot and borax to absorb scattered neutrons and gamma radiations due to neutrons stopped in the steel. The resolution time, determined mainly by the cables and electronic systems employed, was 0.1 μsec, giving $dt/l \sim 0.5$ nsec $-$ m^{-1}, comparable with the best laboratory measurements.

The neutron beam was incident from below on the detector stack shown in Figure 1. We were concerned primarily in this experiment with the determination of fission cross sections. The seven targets, from bottom to top, where Pu^{239}, Pu^{240}, U^{235}, a blank nickel holder, a Li^6 fission foil, Pu^{241}, and Pu^{242}. Camera shutters, shielding the counters from the highly

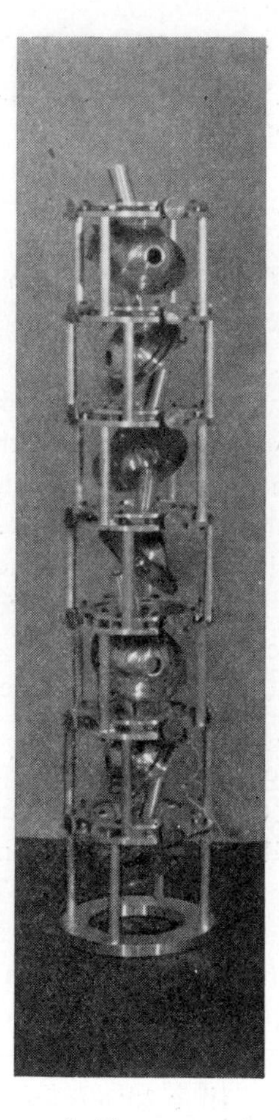

Fig. 1. Photograph of the detector stack. The neutron beam is incident from below. There were seven target foils, each viewed by detectors at three angles.

Fig. 2. Photograph of the amplifier box mounted on the sled. This compartment was air conditioned for use in the desert.

Fig. 3. Photograph of the radioactive foils being mounted in the vacuum chamber on the front of the sled.

Fig. 4. Photograph of the sled in position beneath the instrument tower.

Fig. 5. Photograph of the instrument tower. Its dimensions were 17 × 17 × 50 ft. Various experiments were performed on the different levels in this tower.

Fig. 6. Photograph of the inside of the recording shed. The moving film cameras record the signals from the detectors, displayed on oscilloscopes. Each camera can photograph two or three oscilloscopes, and each oscilloscope can display two signals.

radioactive foils, were opened only two seconds before the explosion. The amplifiers, shown in Figure 2, had a logarithmic characteristic between 10 millivolts and 10 volts, and were linear at lower energies. Thus we could cover essentially four decades of signal. Such a dynamic range is necessary for recording both high energy (early time, higher flux) and low energy data.

In order to be able to recover the targets and electronics, the detector stack and the amplifier compartment were mounted on a sled. The vacuum chamber mounted on the front of the sled is shown in Figure 3, where a radioactive target is being installed. The sled is shown in position in Figure 4, directly over the vacuum pipe from the source underground. Additional experiments were performed in the tower, shown in Figure 5. The signal cables seen here coming out of the tower ran about 1000 ft to recording stations.

Following the explosion, a large bubble of hot gas is formed underground. When this cools, the earth collapses–the fall into this caldera is quite violent: one instant the tower is standing on solid ground, and the next instant

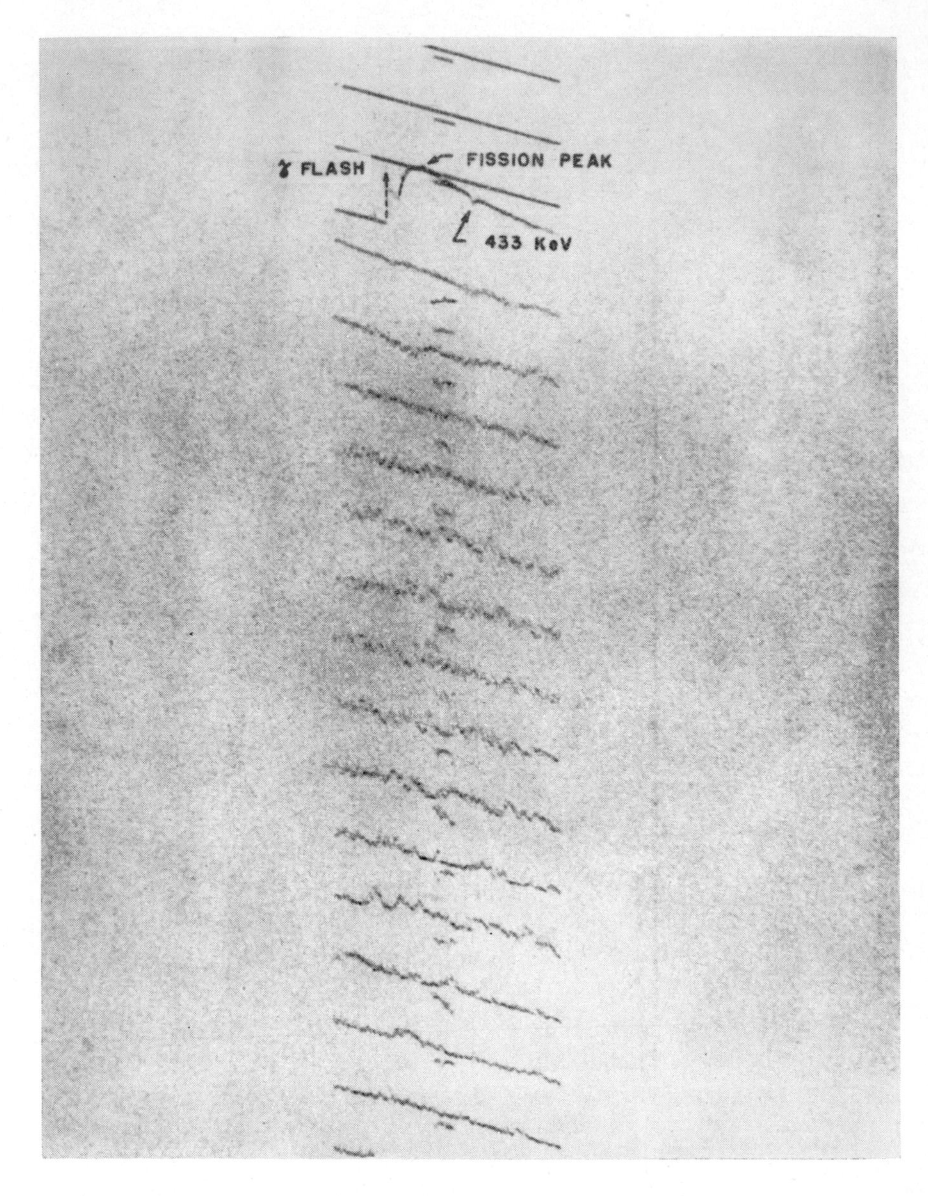

Fig. 7. Oscillogram recorded on 35-mm moving film. The repetitive sweep was 50 μsec; image reduction ratio was 20 : 1; film speed was 100 ft/sec. This is a single beam oscilloscope; most were dual beam.

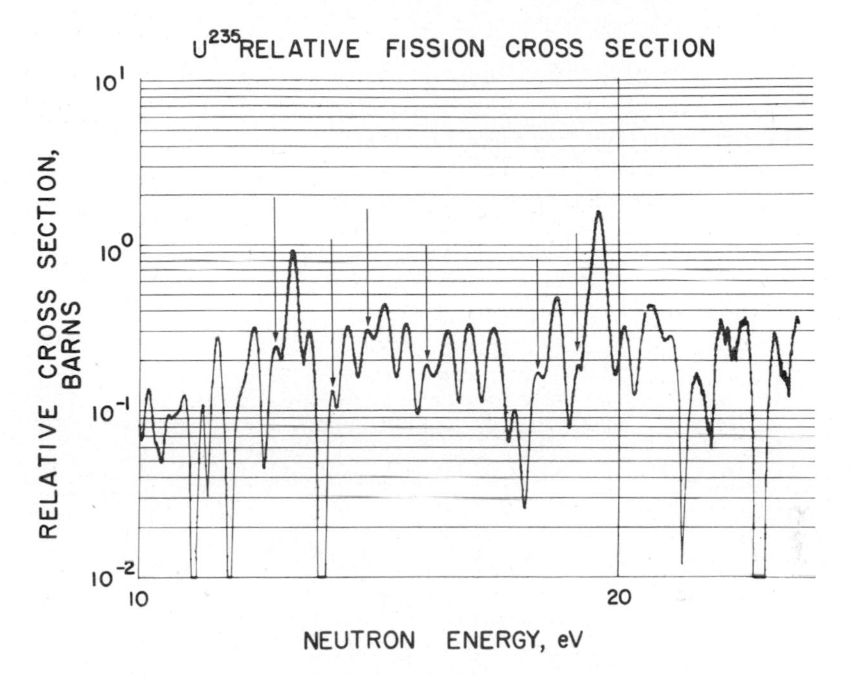

Fig. 8. Fission cross section data for U^{235}, measured by the signal received in millivolts. The arrows indicate "ghost" images on the low energy side of resonances, caused by reflection of neutrons from the bottom of the shaft.

ground level is 20 ft below it. Therefore, as soon as the shot took place, the sled was pulled out of the way. Sled recovery was completely successful. It was roughly three minutes until the crater formed, while only 40 seconds were required for the removal of the sled.

The inside of a recording shed is shown in Figure 6. With the large fluxes realized in this experiment, we did not measure individual events but rather the total current from the detectors. These signals were displayed on oscilloscopes, and recorded continously by the bank of cameras shown in this figure.

A typical example of the camera recordings is shown in Figure 7. Prior to the shot, there are no signals. The continuous oscilloscope traces are seen as straight lines at the top of the figure, slanted by the motion of the film. The short lines, resulting from the reflection of a small portion of the beam by a mirror glued to the oscilloscope face, determined the baseline throughout the signal time. Data continued for about 4 ms, using about 5 inches in the middle of each 10-ft roll of film.

As an indication of the low-energy limit of the data so far obtained, Figure 8 shows some preliminary data for the fission cross section of U^{235} from 10–25 eV. If the flux were known, these results could be converted into relative cross section. It is clear, however, that the resonance structure can be resolved in this energy region. The peaks correspond to well known resonances, but their heights are in error due to poor statistics. In fact, much

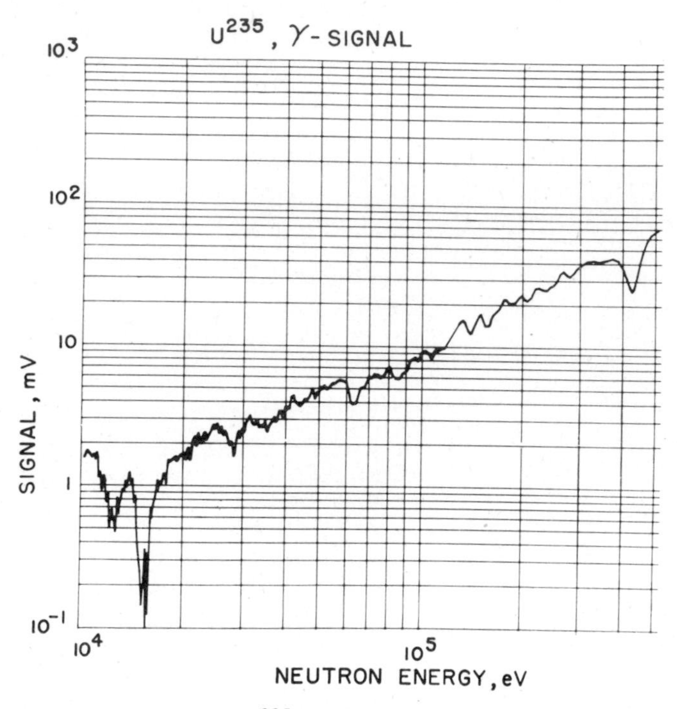

Fig. 9. Gamma ray spectrum for U^{235} target in the energy range 10–500 keV, This includes both fission and capture gammas.

better results in this energy range have since been obtained in a similar experiment in which a moderator was placed at the bottom of the vacuum pipe, greatly increasing the low energy flux.

It is also possible, in such an experiment, to obtain data in the energy range of interest in nucleosynthesis. Figure 9 shows the gamma signal from a U^{235} target located in the tower, in the energy range 10–500 keV. The gamma-ray detectors employed are patterned after those built by Moxon and Rae (1963) at Harwell, which have particularly low sensitivity to neutrons. All of the windows of the vacuum chamber and the foil holders were made of nickel, allowing us to calibrate these data from several strong nickel resonances. The 433 kilovolt resonance in oxygen is also evident in this figure. Generally, we have good statistics and high flux in this energy region, and hope to develop our techniques sufficiently to make measurements of astrophysical (as well as reactor physical) significance.

REFERENCES

Hemmendinger, A. (1965). *Physics Today* **18**, 17.
Moxon, M. C. and Rae, E. R. (1963). *Nucl. Instr. Methods* **24**, 445.

Distribution of Neutron-Source Strengths for the *s*-Process

DONALD D. CLAYTON

Presented at the NASA Conference on Nucleosynthesis
New York, January 1965

POINT OF VIEW

A large amount of work has been expended in the science of nucleosynthesis since Burbidge, Burbidge, Fowler, and Hoyle (1957; B^2FH) attempted the first systematic treatment of the subject. Many details of the perspective of that paper have been expanded, revised, or discarded in the subsequent years. One of the subjects that has expanded markedly is the synthesis of the elements heavier than the iron peak, the so-called *heavy elements*. Substantial elaborations of the mechanisms discussed by B^2FH have been made by A. G. W. Cameron and by two of Fowler's students, P. A. Seeger and myself. This work has been predicated on the point of view that charged particle reactions are mainly responsible for the synthesis of elements up to and including the iron "equilibrium" peak, whereas neutron capture has been called upon to account for the major synthesis between $A = 65$ and $A = 270$ (the limits are not sharp).

The necessity of neutron capture arises because charged-particle lifetimes become excessively long for large values of the nuclear charge Z at the temperatures to be expected in the static phases of stellar evolution, whereas the tendency toward thermodynamic equilibrium at very high temperatures tends to destroy heavy elements rather than synthesize them. The main theme of the B^2FH discussion of the heavy elements, however, was that they show prominent abundance features that are apparently only interpretable in terms of neutron capture mechanisms, which are apparently of two types: (1) rapid (compared to β-decay times) capture of free neutrons in a high-temperature high-density bath (r-process), and (2) slow (compared to β-decay times) capture of free neutrons (s-process). For the latter process, which is the one under discussion in this paper, the confirming abundance evidence is based upon the differential equation governing the abundances of adjacent steps in the chain along the path of beta stability:

$$\frac{dN_A}{dt} = (\text{neutron flux}) \{\sigma_{A-1} M_{A-1} - \sigma_A N_A\}, \tag{1}$$

where σ_A represents the (n, γ) cross-section of species A after averaging over the thermal spectrum of the neutrons. The complete set of differential equations is clearly of a type that attempts to minimize the difference in the two terms on the right hand side, tending to the approximate equality

$$\sigma_A N_A \simeq \sigma_{A-1} N_{A-1}. \tag{2}$$

Equation (2) may be called *the local approximation* for the s-process. It has been the source for the all-important experimental confirmation of the correctness of the s-process idea (see the contribution of Gibbons and Macklin in these proceedings), and for several important applications of the s-process to astrophysical theory. Prominent among the latter are the B^2FH discussion of s-process time scales based upon competition between neutron capture and β-decay, Cameron's (1959) application of excited-state beta decay to that competition, and Clayton's (1964a) application to galactic chronology. All of these important discussions were based upon the local approximation.

The emphasis of this paper is to be on the gross features of the s-process abundance curve rather than on local properties. The major point I want to make is a rather obvious one—namely that once the seed nuclei for the s-process and its astrophysical site have been selected, the gross properties of the s-process abundances are immediately relatable to stellar evolution and galatic evolution. B^2FH limited their discussion to the observation that, although Eq. (2) is not a strict equality, the gross features of the abundances should be such as to present a smooth and monotonic decrease of the product σN as a function of increasing atomic weight. The confirmation of this prediction, in spite of the fact that the cross-sections and the abundances themselves have (anti-correlated) maxima and minima, has removed all reasonable doubt of the basic correctness of the s-process idea. A recent example of the solar-system σN_s curve is shown in Figure 1, wherein attention is here called to the experimental points rather than to the calculated curve which will be described later. The almost smooth and monotonic cecrease of these σN_s values, belonging only to those nuclei independently believed to by synthesized in the *s*-process, bears a marked contrast to the corresponding product for those nuclei independently believed to be due to other nuclear processes. (For example, Figure 2 of CFHZ).

An important conceptual step forward was made by Clayton, Fowler, Hull, and Zimmerman (1961; CFHZ), who began to analyze the neutron fluxes required to convert seed nuclei into the observed distribution of the σN_s products for solar-system abundances. To analyze the fluxes required to produce a desired result obviously requires some assumption regarding the identity of the initial nuclei—those seed nuclei which are to be converted into the heavy elements. CFHZ adopted Fe^{56} as the seed nuclei for their analysis. Because the deduction of the neutron fluxes in stellar interiors is dependent upon the seed nuclei assumed, it is worth reviewing the basic reasons for choosing Fe^{56}.

The abundance peak at Fe^{56} is not synthesized by neutron capture, but rather by the network of charged particle reactions which liberate fusion energy by converting light nuclei to those nuclei at the maximum of the binding-energy/nucleon curve, which maximum occurs at Fe^{56}. B^{2}FH viewed this synthesis as the thermodynamic equilibrium of nuclear matter, which in a significant temperature-density domain favors Fe^{56} as the most abundant nuclear species. The corresponding process has come to be called the *e*-process, even though thermodynamic equilibrium is not actually attained. (See Fowler and Hoyle (1964) and Gilbert et al. in these proceedings). In spite of the present uncertainty regarding this process, it seems certain that the resulting abundance peak around $A = 56$ is much narrower than the observed peak in the σN curve of the *s*-process nuclei. That is, the CuZnGeSeKr drop-off in Figure 1 occurs much too slowly and smoothly to be the result of the *e*-process. So one is led to conclude that these nuclei were synthesized by addition of neutrons to the narrow *e*-process abundance peak, even though the same conclusion cannot be extended to the heavier nuclei. That is, there remains the possibility that the very heavy nuclei were synthesized from some lighter seed nucleus (e.g., Ca^{40}, Si^{28}, *etc.*) Such a possibility must be consistent, however, with the fact that neutron-capture cross-section for nuclei lighter than Fe tend to be much smaller than for those heavier than Fe. This fact has the interesting consequence that it takes less neutron flux to convert Fe to Pb than to convert Si^{28} (say) to Ba. Thus one can conclude that light nuclei could have been the seed for the very heavy nuclei only if little or no irongroup nuclei were present in the bath. The point is significant enough to bear stating in another way. Calculations assuming Fe^{56} to be the seed will show that the probability of finding a large integrated neutron flux decreases so rapidly with increasing flux that a flux distribution capable of converting Fe^{56} to the observed s-process abundances is inadequate to convert light elements into heavy elements. In light of the foregoing two arguments, it is certainly reasonable to assume that the iron e-peak served as a seed for *all* the heavy s-process nuclei. Since the abundance of Fe^{56} dominates that abundance peak, moreover, the simplest realistic analysis is to use the unique seed nucleus, Fe^{56}.

CFHZ analyzed the addition of neutrons at constant temperature in terms of the time-integrated neutron flux

$$\tau = 10^{-27} \int_0^t v_T n_n(t)\, dt, \tag{3}$$

where $v_T = \sqrt{2kT/m_n}$ and n_n is the neutron density. The factor 10^{-27} is introduced in order that Equation (1) becomes

$$\frac{dN_A}{d\tau} = \sigma_{A-1} N_{A-1} - \sigma_A N_A, \tag{4}$$

when the thermally averaged cross sections are expressed in the common units of millibarns (10^{-27} cm^2). If the σN product *per initial* Fe^{56} *nucleus* resulting from its exposure to an integrated neutron flux equal to τ is defined as

$$\psi_k(\tau) \equiv \sigma_k N_k(\tau)/\mathrm{Fe}^{56}(\mathrm{O}), \tag{5}$$

where the new index is $k = A - 55$, the exact solution to the abundance distribution is

$$\psi_k(\tau) = \sum_{i=l}^{k} C_{ki} \mathrm{e}^{-\sigma_i \tau}, \tag{6}$$

where

$$C_{ki} = \frac{\sigma_1 \sigma_2 \sigma_3 \ldots . \sigma_k}{(\sigma_k - \sigma_i)(\sigma_{k-1} - \sigma_i) \ldots . (\sigma_1 - \sigma_i)}, \tag{7}$$

$$\text{omitting} \frac{1}{(\sigma_i - \sigma_i)} .$$

An alternative approximate solution of good accuracy (CFHZ) for most purposes at the present state of knowledge is

$$\psi_k(\tau) \simeq \lambda_k \frac{(\lambda_k \tau)^{m_k - 1}}{\Gamma(m_k)} e^{-\lambda_k \tau}, \tag{8}$$

where the parameters λ_k and m_k are determined by the cross sections according to

$$\lambda_k = \frac{\sum_{i=1}^{k} \frac{1}{\sigma_i}}{\sum_{i=1}^{k} \frac{1}{\sigma_i^2}} = \frac{\left\langle \frac{1}{\sigma} \right\rangle_k}{\left\langle \frac{1}{\sigma^2} \right\rangle_k}, \tag{9}$$

and $$m_k = \frac{\left(\sum_{i=1}^{k} \frac{1}{\sigma_i} \right)^2}{\sum_{i=1}^{k} \frac{1}{\sigma_i^2}} = \frac{k \left\langle \frac{1}{\sigma} \right\rangle_k^2}{\left\langle \frac{1}{\sigma^2} \right\rangle_k}. \tag{10}$$

The brackets $\langle \quad \rangle_k$ designate the average of the first k terms. This approximate solution is very useful in the analysis of the gross properties of the σN curve. It is accurate to 20% for almost all values of k and τ and is easier to use for certain simple superpositions of neutron fluxes. CFHZ were able to show (see Figure 14 of their paper) that the solar-system σN curve could not be the result of a single neutron exposure of iron nuclei. If the basic point of view is correct, the solarsystem abundances must reflect a superposition of curves corresponding to differing neutron fluxes, each contribution having an appropriate normalization reflecting the number of iron nuclei exposed to that flux. Specifically, if $\varrho(\tau)\, d\tau$ is the number of iron nuclei exposed to a neutron flux τ in the range $d\tau$, the resultant abundances are given by

$$\sigma_A N_A = \int_0^\infty \varrho(\tau)\, \psi_A(\tau)\, d\tau. \tag{11}$$

A natural consequence of Eq. (11) is the near constancy of the composite σN values between magic-neutron shells, since each Ψ_A curve is itself nearly constant between shells. The major changes in the value of σN should occur *at* the closed neutron shells. These features are presently evident in Figure 1, which shows a ledge-precipice structure near $A = 90$ and $A = 140$.

The discussion leading up to Eq. (11) defines a point of view toward the *s*-process that will be used in the subsequent analysis. It is important that the basis of this point of view be periodically criticized, because it provides, if it is correct, a powerful tool for linking stellar and galactic evolution to heavy element abundances. Let me emphasize that the *s*-process mechanism itself is confirmed beyond reasonable doubt. It is in the assumption of the seed nuclei and astrophysical circumstances that some uncertainty remains,

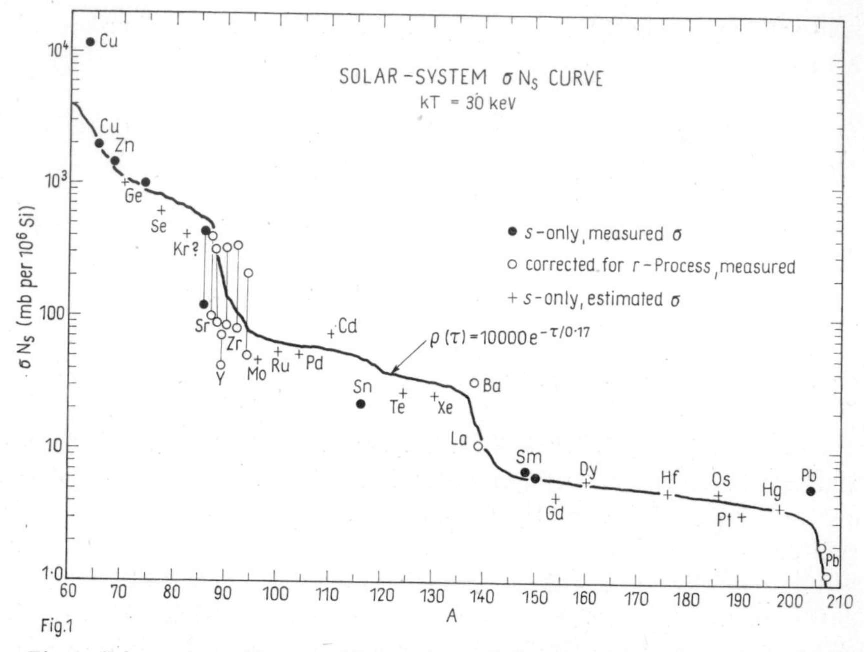

Fig. 1. Solar-system σN_s curve. The product of the neutron-capture cross section at $kT = 30$ keV times isotopic abundance is plotted versus atomic mass number A. The solid line is a calculated curve corresponding to an exponential distribution of integrated neutron flux.

and both of these assumptions are germain to the extraction of the history of neutron fluxes in stellar interiors.

At least in stars later than second generation one expects elements heavier than iron in the *initial* composition of the star. No ambiguity in the integrated neutron flux results if these are *s*-process heavy elements, because the variable τ is cumulative and need not be accrued within one star. One simply considers all the iron, whenever formed, and traces it through galactic history to the present time. Then one asks today (or 4.5×10^9 years ago if solar abundances are to be analyzed) for the spectrum of accumulated neutron exposures to which the total amount of iron produced was exposed. In this sense, for example, one must remain alert to the possibility that the maximum exposures required are the results of the small amount of iron *s*-products that have been processed through two or more generations.

There is another source of heavy elements, the *r*-process nuclei (see P. A. Seeger's contribution in these proceedings). It is no trivial matter to evaluate the extent to which the *r*-process elements may have served partially as seed for the production of *s*-nuclei, particularly since very small fluxes suffice to convert the Te-Xe peak to Ba and the Os-Pt peak to Pb. The simple but not completely conclusive argument for discarding this seed is that probably no more than a few percent of all the *r*-nuclei have ever been reprocessed through an *s*-process event. My attitude is that one should first attempt to analyze the *s*-nuclei without this complication and follow by a self-consistency analysis with the derived $\varrho(\tau)$ in which one assumes that the flux spectrum for the reprocessed *r*-nuclei is the same as the spectrum for the processed iron peak. I will return to these points in the concluding section.

NEUTRON SOURCES AND TEMPERATURE

Equation (11) is oversimplified in one very important way; *viz.*, it rests on the assumption that all of the *s*-processing occurred at the same temperature. Because the neutron-capture cross sections of the heavy elements are temperature dependent (see Macklin and Gibbons, 1965, and Gibbons in these proceedings), $\psi_A(\tau)$ is not really a function only of the neutron flux τ unless the discussion is restricted to constant temperature. At higher temperatures, the cross-sections are smaller; *i.e.*, it would require a larger neutron flux τ to process the initial Fe^{56} to the same result. This problem is not as serious as it seems, however, for it will be seen shortly that the parameter τ can be temperature-scaled in a way that makes all τ's comparable.

Although there are many fine points to be considered regarding the neutron sources, it seems to me that there are three primary requirements to be met by any adequate source:

(1) It must be capable of liberating considerably more neutrons (say by a factor of 10) than there are heavy seed nuclei to adsorb them.

(2) There must not be so many light neutron poisons that the heavy elements are unable to compete for the neutrons.

(3) The source must occur in an astrophysical site that can reasonably be ejected into the interstellar medium without a disruption of the nuclear constitution.

The first two of these requirements have been discussed often, but the third requirement seems hardly to have been appreciated at all. But, for example, any *s*-processing during helium or carbon-oxygen burning at the center of a star is irrelevant unless that core can, at least in part, be expelled without encountering the usual terminal phases, either catastrophic or degenerate. It has been a scientific mismatch that for nucleosynthesis the terminal stages of stellar evolution and mass loss are the overriding issues, whereas the science of stellar evolution has been progressing from the initial static phases.

In spite of many ingenious attempts, only three neutron sources for the *s*-process are commonly discussed. I want only to comment on each of the three sources with emphasis on two features—that the temperature domains of the three sources differ and that the efficiency of each source will be continuously variable.

(A) C^{13} (α, n) O^{16}, and in the same way O^{17} (α, n) Ne^{20}. Greenstein's (1954) suggestion of this source was initially interpreted in terms of residual products of the CNO cycle being gravitationally heated in a helium bath following hydrogen exhaustion. This source, however, fails both in requirements (1) and (2) listed above. The source has been revived by Caughlan and Fowler (1964), however, in terms of a model that mixes protons (resulting in the production of C^{13}) into a helium-burning zone (they envisioned the core of a red giant). It seems almost certain to me that if this mechanism is operative, its efficiency will vary continuously with mass and composition of the star involved. The temperature might range all the way from $T_8 = 1$ to $T_8 = 3$.

(B) Ne^{22} (α, n) Mg^{25}. First suggested by A. G. W. Cameron, this source has the attractive feature that Ne^{22} may be very abundant (from two (α, γ) reactions on N^{14}] in heated hydrogen exhausted zones that formerly operated on the CNO cycle, and that the neutron poison problem may not be too severe. The critical feature of this neutron source is its temperature sensitivity. The (α, γ) channel dominates the (α, n) channel below $T_8 = 2.1$, whereas the (α, n) channel dominates above $T_8 = 2.1$. It has by no means been ascertained whether a sufficient amount of helium can survive to the high temperatures $T_8 \geqq 2.0$ necessary to liberate neutrons, although Iben (1965; see contribution in these proceedings) has expressed a negative result for the heliumburning core at $3M_\odot$. But the lifetime of Ne^{22} varies as T_8^{-27} near the crossover at $T_8 = 2.1$, at which point its value may be about 10^3 yr. if a nominal amount of helium has survived. It is, however, very clear that, if this source is ever effective, its yield must vary markedly with composition and mass of the host star. In fact, the temperature gradient in radiative zones should cause the neutron yield to vary with radius. Most of the neutrons liberated from this source will be captured at $T_8 = 2.0$ to 2.5.

(C) Carbon-Oxygen burning. This neutron source seems the most difficult to analyze. In particular, the restriction of requirement (3) of the neutron sources may eliminate it as a *core* source, in which case it would have to operate in a shell during the short-time-scale stages when the core is even more evolved. At any rate, the neutron yield should be a continuous function of the stellar variables and should occur mostly in the range $T_8 = 6 - 8$.

COMPOSITE σN CURVES

Most of the analyses of the *s*-process σN curve have been made for temperatures near 30 keV because the best cross-section data have been available near that energy and because it represents as good a compromise as any between the anticipated temperatures of the probable sources. The points of Figure 1, which employed 30 keV cross sections, could be adequately fit by a discrete superposition of 30 keV exposures; *viz.*,

$$\sigma_A N_A = 2160\,\psi_A(\tau = 0.1) + 990\,\psi_A(\tau = 0.2) + 45\,\psi_A(\tau = 0.6) + 3.6\,\psi_A(\tau = 1.1). \tag{12}$$

The coefficients in Eq. (12) represent the number of Fe^{56} atoms per 10^6 Si atoms exposed to the flux τ. Although other superpositions of the form of Eq. (12) would work just as well, they share the property that the number of seed exposed to the integrated flux τ is a rapidly decreasing function of τ. Since each neutron source is expected to produce an integrated flux that varies continuously with stellar mass, composition, and radial point, it seems to me that the distribution of exposures should be continuous; that is, $\varrho(\tau)$ should be a continuous function of τ rather than a series of four delta functions. Seeger, Fowler, and Clayton (1965; SFC) have presented the first analyses of that assumption. If $\varrho(\tau)$ has the form

$$\varrho(\tau) = Ge^{-\tau/\tau_0}, \tag{13}$$

where G and τ_0 are constants to be selected, Eq. (11) can be integrated using the approximate (also the exact) form of $\psi_A(\tau)$ to give

$$\sigma_A N_A \simeq G\left[\frac{\lambda_A \tau_0}{\lambda_A \tau_0 + 1}\right]^{m_A} \tag{14}$$

where λ_A and m_A are given by Eq. (9) and (10). SFC presented graphically the solutions for a spectrum of values of τ_0, and the particular case with $G = 10^4$ and $\tau_0 = 0.17$ is shown as the calculated curve in Figure 1.

Figure 1 shows that the simplest physical assumption [a smooth and monotonic $\varrho(\tau)$] results in a structured σN curve. It is the increasing definition of the σN curve with improved abundance and cross-section measurements that is making possible the extraction of the astrophysically significant quantity, $\varrho(\tau)$.

As a clarifying example let me say at this point what modifications would be required if the *s*-process occurs at a single temperature which is different than $kT = 30$ keV. First the points of Figure 1 must be replotted with the new cross sections—the major effect would appear as a slight shift in the normalization of the curve. Then the new points would be fitted to a curve calculated from cross sections at the appropriate temperature. SFC displayed sets of such curves for $kT = 15$ keV and $kT = 55$ keV. The shape of the family of curves is essentially unchanged, but the value of τ_0 (which physically measures the fluxes) appropriate to a given shape changes with temperature. The near identity of the shapes means that the shape of the observed σN curve is not an indicator of the temperature of the *s*-process. It can only yield the flux distribution once the temperature has been independently selected.

If more than one temperature has participated in the *s*-process it becomes necessary to decompose $\varrho(\tau)$ into its temperature components. This possibility (likelihood) is complicated by the fact that the processing may be cumulative, and *I* would distinguish basically two types of cumulative processing: (1) heavy nuclei irradiated in one phase of stellar evolution (say He-burning) could be irradiated further in a later phase (say carbon burning) of the same star, and (2) nuclei irradiated in one star may be ejected to the interstellar medium, reincorporated into another star, and irradiated further In both cases the integrated neutron flux is cumulative but complicated by the possibility that the temperatures of the phases differ. Without attempting to treat this situation exactly (the complication is unwarranted at present) let me point out how the simple analysis outlined here may be approximately preserved.

It is fortunate (in one sense) that most of the neutron cross sections scale by roughly the same factor with temperature change. It will be obvious from the exact solution in Eq. (6) that if all the cross sections at temperature T are related to those at 30 keV by a constant conversion factor, i.e.,

$$\sigma_A(kT) = r(T) \quad \sigma_A(30),$$

then the abundance distribution from a flux τ_T at temperature T would be exactly equal to the abundance distribution from a flux at $kT = 30$ equal to $\tau_{30} = r(T)\ \tau_T$. What SFC showed for the actual cross sections is that the gross distribution for irradiation at kT is approximately equal to that for irradiation at 30 keV if the integrated fluxes are related by

$$\tau_T \simeq \left[\frac{kT}{30}\right]^{0.7} \tau_{30}. \tag{15}$$

As far as gross properties are concerned, the average effective cross sections scale as $T^{-0.7}$. To reasonable approximation, therefore, Eq. (11) can be retained and approximate analysis of the gross properties of the composite curve can be made *using* 30 *keV cross sections provided* τ *is reinterpreted as a measure of*

$$\tau = \int v_{T(t)} \left\{\frac{30}{kT(t)}\right\}^{0.7} n_n(t)\,dt. \tag{16}$$

The temperature dependence of this quantity is small since $v_T \propto T^{0.5}$—in fact, if all the cross sections had a $1/v$ dependence, it is easy to show that the new τ would be simply the velocity at 30 keV multiplied by the time integral of the neutron density, i.e., no temperature dependence.

In summary then, first-order analysis can be performed by extracting $\varrho(\tau)$ with the use of Eq. (11) and 30 keV cross sections. The variable τ is to be interpreted as the time integral of the neutron flux times a fudge-factor, $(30/kT)^{0.7}$, which approximately measures the way thermal neutron cross sections scale near $kT = 30$ keV. The use of Eq. (16) allows a cumulative measure of τ when the irradiation has not all been at the same temperature.

A SPECULATIVE EXAMPLE

As evidence of the fact that the ledge-precipice structure of the curve in Figure 1 is not a special consequence of the choice of an exponential form for $\varrho(\tau)$, SFC obtained the same general type of curve for $\varrho(\tau) = G\tau^{-n}$, which they also integrated explicitly. The point to be drawn from these calculations is that any monotonically decreasing $\varrho(\tau)$ will produce curves of the same general structure. That this structure appears to be in line with the observed facts lends strength to the point of view adopted. In fairness to the points of Figure 1, however, it is apparent that a smoothly decreasing curve between Cu and Zr would fit the points as well as, if not better than, the structured ledge and precipice of the computed curves. I have recently (Clayton, 1964b) called attention to this problem and to the central role of the geochemistry of Sr Y Zr in its resolution. But as an example of the techniques of this paper, let me make a speculative interpretation of the possibility that drop-off is smooth.

If the σN curve is not to have the structure associated with a smooth $\varrho(\tau)$, then $\varrho(\tau)$ must possess a compensating structure. The original analysis of CFHZ shows clearly what that compensating structure would have to be, for the ledge-precipice structure near $A = 90$ is primarily due to integrated neutron fluxes near $\tau = 0.3$. The lower-left-hand corner of Figure 2 shows

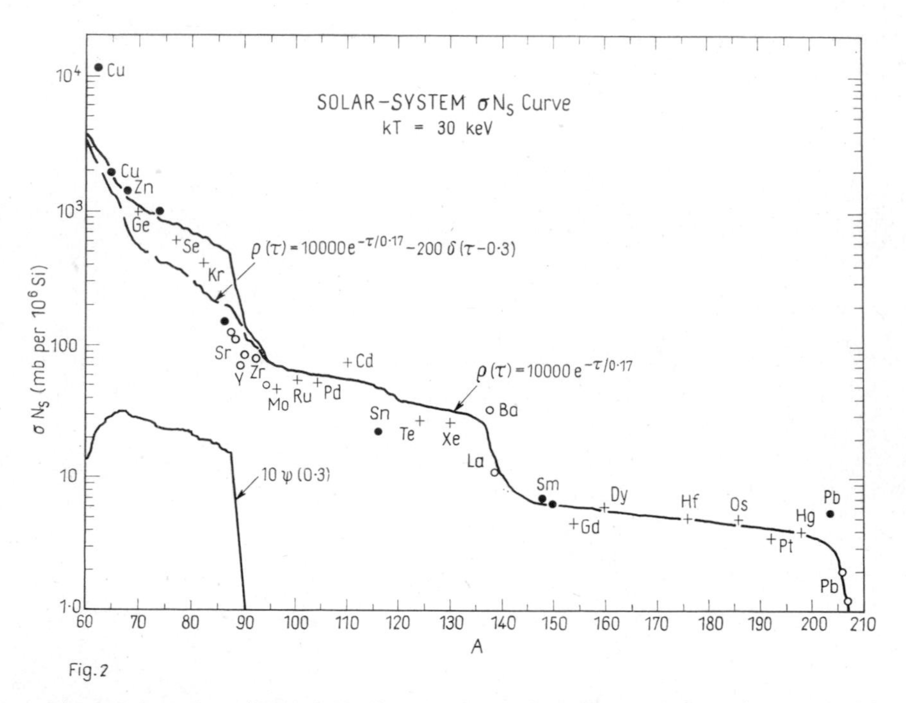

Fig. 2. Subtraction of 200ψ (0.3). The subtraction of $200\delta(\tau - 0.3)$ from the exponential $\varrho(\tau)$ produces a calculated σN_s curve that decreases more smoothly than that of Figure 1 between $70 < A < 110$. The resulting dashed curve could be moved through the points by raising $\varrho(\tau)$ for small values of τ. Further geochemical research is required to establish he extent of the precipice near $A = 90$.

the curve ψ (0.3). The way to eliminate the ledge from the composite σN curve is to have a $\varrho(\tau)$ with a deep minimum near $\tau = 0.3$. Also shown in Figure 2 is the curve of Figure 1 and the result of subtracting $200\,\psi(0.3)$ from that curve. It can be seen that this curve does decrease more smoothly. Keep in mind, too, that the left end of the curve should be raised because the exponential $\varrho(\tau)$ is not large enough for the small values of τ. A negative delta function in $\varrho(\tau)$, is, of course, not physically interpretable, but essentially the same result is obtained from

$$\varrho(\tau) = 10^4 e^{-\tau/0.17} \qquad \tau = < 0.25,\ \tau > 0.36$$

$$\varrho(\tau) = 0 \qquad 0.25 < \tau < 0.36.$$

This function is still not reasonable, but it does illustrate that $\varrho(\tau)$ would have to have a very deep minimum near τ = p.3 to obtain a smooth drop. To bring the left side back up, the real $\varrho(\tau)$ would have to have a shape like the one shown schematically in Figure 3.

If (and I stress *if*) $\varrho(\tau)$ does in fact have these features, what would be their likely implications for the galactic history of solar-system material? I would interpret a double peak in $\varrho(\tau)$ as signifying that there have been at least two astrophysical neutron sources of importance for the synthesis of the heavy nuclei—one which produced fluxes predominantly smaller than

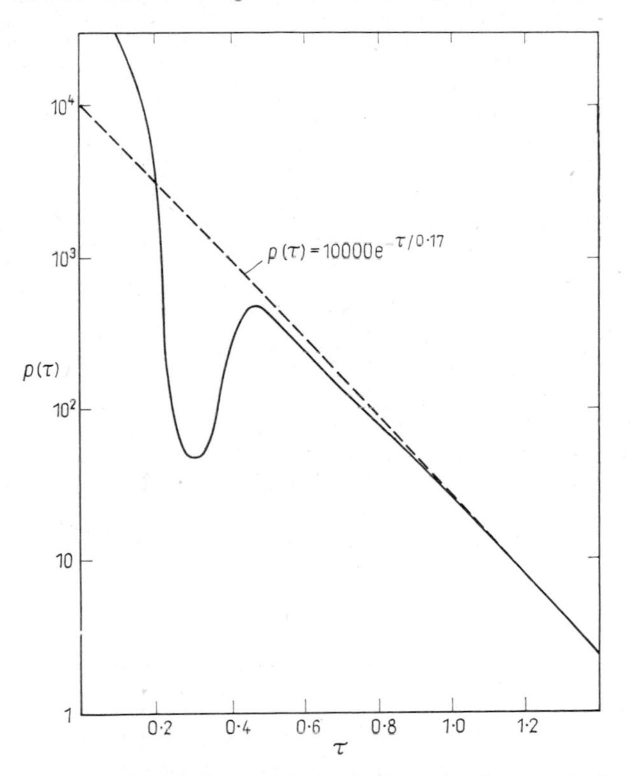

Fig. 3. A doubly-peaked distribution of integrated neutron fluxes is required to minimize the precipice in σN_s near $A = 90$. For small values of τ, $\varrho(\tau)$ should exceed the exponential value to fit the Cu Zn Ge points (see Figure 2). This flux distribution, if it is required, would be interpreted as evidence of two physically distinct sources of *s*-process neutrons.

$\tau = 0.2$, and another which produced a spread in fluxes about a most probable value $\tau = 0.5$. I would *tentatively* interpret the larger amount of iron seed exposed to the smaller fluxes as being remnants of helium-burning phases and the smaller amount of iron seed exposed to the larger fluxes as being remnants of carbon burning. The fact that the direct remnants of carbon burning (Mg, Si, S) are an order of magnitude less abundant than the direct remnants of helium burning (*C*, *O*) certainly suggests that the amount of iron exposed to neutron fluxes in the former is also less than that in the latter.

If this, or any other, identification of component sources in $\varrho(\tau)$ can be made, the 30-keV $\varrho(\tau)$ can then be reinterpreted in terms of real flux histories. For example, if the hypothetical discussion of this section should turn out to be true, the two peaks in Figure 3 need to be shifted to obtain the true integrated flux distributions. If the first component is due to Ne^{22} (α, n) Mg^{25}, which would imply $kT \simeq 20$ keV, then it will be contracted in terms of true neutron flux according to Eq. (15):

$$\tau_{20} = \left\{\frac{20}{30}\right\}^{0.7} \tau .$$

If the second peak is due to carbon burning, which would suggest $kT \simeq$ 55 keV, then it will be expanded in terms of the actual integrated neutron fluxes:

$$\tau_{55} = \left\{\frac{55}{30}\right\}^{0.7} \tau .$$

It would then remain to be seen whether the assumed astrophysical environments are capable of producing these two components of the integrated neutron-flux distribution. I think there is no reason today to believe that the speculations of this section are correct, but they illustrate the potential power of this type of analysis.

CONCLUSIONS

A good caricature of the spectrum of integrated neutron fluxes to which iron has been exposed would seem to be $\varrho(\tau) = 10^4 e^{-\tau/0.17}$ per 10^6 Si atoms. This function is not great enough for $\tau < 0.1$, and it may have to be amended by a minimum near $\tau = 0.3$ if the drop near $A = 90$ is smooth. But for the sake of one last bit of discussion I will assume that the exponential of Figure 1 is reasonable for $\tau \geqq 0.1$. Not only should this function provide a boundary condition for models of the nuclear evolution of the galaxy, but it also brings up several interesting points regarding the self-consistency of the analysis. I will comment on a few of these.

The solar abundance of Fe is $1.2 - 2.0 \times 10^5$ (Goldberg, Müller, and Aller, 1960, and Goldberg, 1964), whereas the chondritic values seem to cluster about 8×10^5 (Urey, 1964). It is important to know this abundance not only for the cosmogonical problem but also to be able to extract the percentage of Fe nuclei that have been exposed to a given neutron flux. In the following I will take the published solar value $\text{Fe} = 1.2 \times 10^5$. (The smaller the Fe abundance, the greater is the percentage that has been through the s-process). Then the fraction of iron nuclei exposed to fluxes greater than $\tau = 0.1$ is

$$\frac{\int_{0.1}^{\infty} \varrho(\tau)\, d\tau}{\text{Fe}} \simeq \frac{920}{1.2 \times 10^5} = 0.76\%. \tag{17}$$

Now the end product of the s-process is Pb, and CFHZ showed that fluxes $\tau \geqq 1.8$ suffice to convert almost all the Fe to Pb. In light of the steepness of $\varrho(\tau)$, we may wonder whether the small amount of seed exposed to such large fluxes, may not be due to the cumulative irradiations rather than single-generation irradiations. The answer to this question has great significance with regard to the s-process capabilities to be demanded from a single star. That such an explanation of large exposures is possible can be seen from the ratio of iron nuclei exposed to fluxes capable of making Pb to the number of iron nuclei exposed to fluxes half that size:

$$\frac{\varrho(1.8)}{\varrho(0.9)} = e^{-0.9/0.17} \simeq 0.5 \times 10^{-2}. \tag{18}$$

That this number is almost equal to the fraction of Fe nuclei exposed to neutron fluxes (i.e., remixed) confirms the point. The terminal exposures of the s-process may be very involved indeed.

The entire question of Pb can be reexamined with a knowledge of $\varrho(\tau)$. One requirement of self-consistency is that $\varrho(\tau)$ have an *asymptotic form* consistent with the s-process abundance of Pb:

$$A = \sum_{206}^{208} \int_{0}^{\infty} \varrho(\tau) \frac{\psi_A(\tau)}{\sigma_A}\, d\tau = \text{Pb}_s, \tag{19}$$

where Pb_s is the lead abundance due to the s-process.
The functions ψ_A are not those given in Eq. (6)–(9) because the solutions are complicated for Pb because of recycling from α-decay. I have obtained the solutions and will publish an analysis of Pb when the data are ripe. (Clayton

and Rassbach, 1967). The problem has added importance for galactic chronology (Clayton, 1964a). In that analysis I used equilibrium abundances (the local approximation) for the Pb isotopes, but I will state at this time that that approximation is unwarranted for the trial extraction of $\varrho(\tau)$.

Fluxes less than $\tau = 0.05$ are sufficient to convert the $N = 92$ and $N = 126$ r-process peaks to Ba and Pb respectively. Since the $\varrho(\tau)$ extracted here is not valid for such small values of τ, it is useless for the analysis of the possibility that r-process nuclei have served as seed for th s-process. Suffice it to say here that this possibility is justifiably neglected unless

$$\int_{0.01}^{0.1} \varrho(\tau)\, d\tau/\mathrm{Fe} \geqq 0.1. \tag{20}$$

It seems unlikely that $\varrho(\tau)$ will be that large for small τ but I am unable to establish a limit at the present time.

I have stated the more-or-less "classical" point of view toward the s-process that I have developed with W. A. Fowler. Like all assumptions, this one needs continuous testing and evaluation. But the s-process itself works. It remains for analyses like this one to really marry the s-process to stellar and galactic evolution.

This work was supported by Grant No. AFOSR 855–65 from the Air Force Office of Scientific Research.

REFERENCES

Burbidge, E. M., Burbidge, G. R., Fowler, W. A. and Hoyle, F. (1957). *Rev. Modern Phys.* **29**, 547, Called B^{2}FH.

Cameron, A. G. W. (1959). *Astrophys. J.* **130**, 452.

Caughlan, G. R. and Fowler, W. A. (1964). *Astrophys. J.* **139**, 1180.

Clayton, D. D., Fowler, W. A., Hull, T. E. and Zimmerman, B. A. (1961). *Ann. of Phys.* **12**, 331. Called CFHZ.

Clayton, D. D. (1964a). *Astrophys. J.* **139**, 637.

Clayton, D. D. (1964b). *J. Geophys. Res.* **69**, 5081.

Clayton, D. D. and Rassbach, M. E. (1967). *Astrophys. J.* **148**, 69.

Fowler, W. A. and Hoyle, F. (1964). *Astrophys. J. Suppl.* **91**, IX, 201.

Goldberg, L., Müller, E. A. and Aller, L. H. (1960). *Astrophys. J. Suppl.* **5**, 1.

Goldberg, L., paper presented November 1963 at NASA Conference on Stellar Evolution, New York.

Greenstein, J. L. (1954). *Modern Physics for the Engineer*, Chap. 10, L. N. Ridenour, ed., McGraw-Hill (New York).

Iben, I. (1965). *Astrophys. J.* **142**, 1447.

Macklin, R. L. and Gibbons, J. H. (1965). *Rev. Modern Phys.* **37**, 166.

Seeger, P. A., Fowler, W. A. and Clayton, D. D. (1965). *Astrophys. J. Suppl.* **97**, XI, 121. Called SFC.

Urey, H. C. (1964). *Rev. Geophys.* **2**, 1.

Calculation of *r*-Process Abundances

PHILIP A. SEEGER,
WILLIAM A. FOWLER and
DONALD D. CLAYTON

The purpose of this paper is to give a reappraisal of the means by which the *r*-process elements have been built up. This has been prompted by the appearance of new experimental data on abundances and neutron cross sections, and by an investigation of the systematics of nuclear binding energies done by one of us (PAS). (This has subsequently appeared as a paper—see Seeger *et al.*, 1965).

For a starting point in our computation of the observed abundances of the heavy (heavier than iron) *r*-process elements we have chosen abundances from the range of observed values reported by Urey (1964). The *r*-process abundances can then be determined by subtracting out of the total abundance curve all contributions from the *s*-process.

The latter requires the construction of a suitable σN_s curve to fit solar system abundances, where σN_s, resultant from Fe^{56} as a seed nucleus, should be a smooth curve representable in terms of a superposition of differing neutron exposure histories as

$$\sigma N_s = \int_0^\infty \varrho(\tau)\, \psi(\tau)\, d\tau\,.$$

Here, τ is the integrated product of neutron flux times time, $\varrho(\tau)\, d\tau$ is the number of iron nuclei exposed to integrated neutron fluxes between τ and $\tau + d\tau$, and $\psi(\tau)$ is the σN_s product resulting from the exposure of Fe^{56} nuclei for a given flux history. By the subscript s on N_s we mean only those nuclear species which lie on the *s*-process path and are shielded from the *r*-process, or else have *r*-process contributions small enough that they can be subtracted with reasonable confidence to obtain good estimates of *s*-process abundances.

The cross sections used in the determination of σN_s and, in turn, N_s were obtained from the recent compilation of Gibbons and Macklin (1964) in conjunction with data listed by Clayton *et al.* (1961). Maxwellian cross section averages appropriate to a typical temperature of $kT \approx 30$ keV were used. The effect of temperature variation on the results of the abundance calculation is in general small, because of the similar variations of all isotopic cross sections with temperature.

The *r*-process abundances obtained in the above fashion are shown in Figure 1, where four kinds of data are plotted on the basis of relative contributions of the *s*-process to the total abundance. If the *r*-process contributes less than 20 per cent of the total for a given isotope then it is not plotted

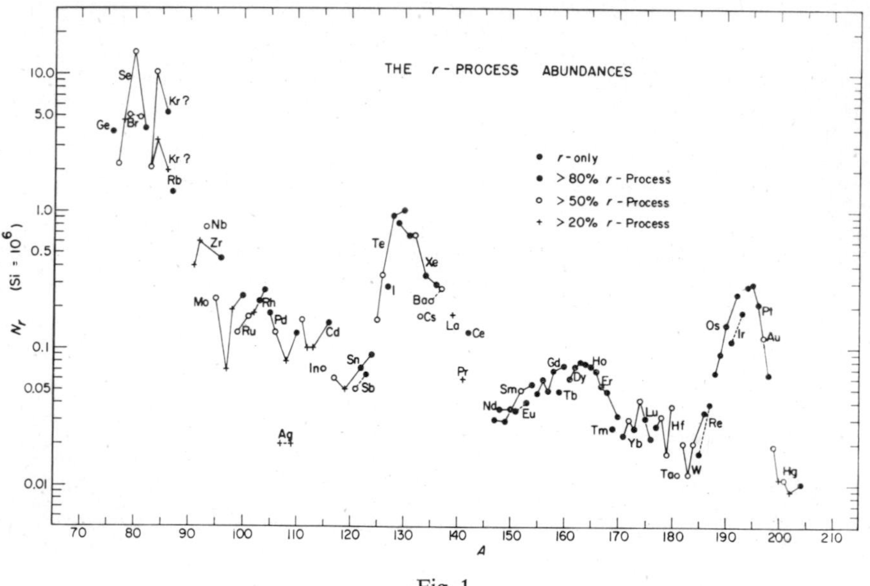

Fig. 1.

because of uncertainties involved in the *s*-process subtraction. Isotopes of the same element are connected in the figure. The main features are the abundance peaks centered about $A = 80$, 130, and 195 which are the same as those that led B^2FH (1957) to formulate the *r*-process originally. Though the existence of these peaks is certain, their relative heights and details are subject to some doubt. For example, the peak at $A = 80$ shown in Figure 1 is very uncertain on the Kr side due to lack of precise knowledge of both *s*- and *r*-process nuclear parameters and time scales. The peak at $A = 130$, seems to be well established except for some uncertainty in height. The abundances in the last peak at $A = 195$ are on a firmer basis than those in either of the first two even though some discrepancies exist between observed meteoritic and solar abundances. No attempt was made to estimate the r-process contributions to Pb and Bi because of the complications involved (see Clayton, 1964), so that calculations were terminated at Hg.

Two other features to be pointed out relate to the structures between the main peaks. The first is the gradual decline in abundance between $A = 80$ and 120 with some indication of minor peak around $A = 100$. If the latter were ascribable to fission and represented the lower hump of a fission distribution, then the upper hump would be hidden by s-process elements and would be neither calculable nor plottable. In any case, this minor peak may represent the effects of recycling of the r-process elements *via* fission. The second feature is the well defined hump centered about the rare earths. It is tempting to attribute this hump to deformations in this region as suggested by Becker and Fowler (1959). In fact, this hump is reproducible qualitatively in our calculations only if deformation terms are introduced in the semi-empirical mass formula.

In order to attribute the abundance peaks at $A = 80$, 130, and 195 to nuclear shell structure, an r-process environment of high neutron number density and high temperature, eg., 10^{24} neutrons per cm^3 and $T = 10^9$ °K is required. Under such conditions (n, ν) and (γ, n) reactions will be rapid enough compared to beta decay rates so that nuclear statistical equilibrium is readily established between nuclei, neutrons and photons. For a given Z, and neutron number density n_n the relative abundances of isotopes A and $A + 1$ at temperature T_9 (units of 10^9 °K) is given by

$$\log n(Z, A + 1)/n(Z, A) = \log \omega(Z, A + 1)/\omega(Z, A) + \log n_n - 34.07 \\ - \tfrac{3}{2} \log \tau_9 + 5.04 Q_n(Z, A)\, T_9,$$

where $Q_n(Z, A)$ is the neutron binding energy, and $\omega(Z, A)$ is the partition function. For these calculations ω has been approximated as being proportional to the number of states within about 1/4 MeV above the ground state, so that

$$\omega \sim 1 + 0.25/\Delta(Z, A)$$

where $\Delta(Z, A)$ is the low lying level spacing. The Δ's were estimated eyeball fits of level densities of the lowest levels of nuclei with even or odd neutron and proton numbers.

Nuclear binding energies were derived from a semi-empirical mass formula of the standard form with shell effects represented by a parabolic term between each pair of magic numbers. To account for the broad hump

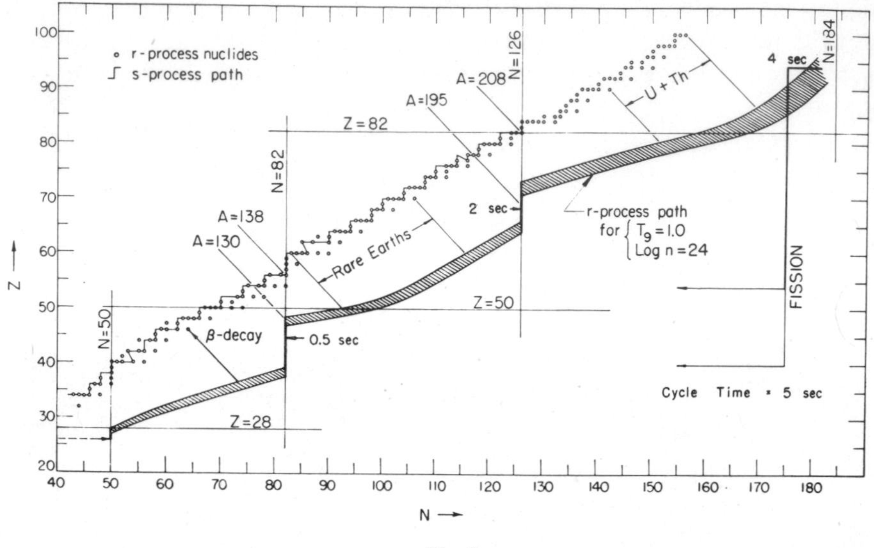

Fig. 2.

in the abundance curve aroung $A = 164$ it was necessary to include deformation terms of the form introduced by Kümmel *et al.* (1964). The whole expression then contains eighteen adjustable parameters which are fit by a least squares adjustment to observed masses. The resultant standard deviation from known masses turns out to be only 0.85 MeV, which is quite good for an 18-parameter formula. Of course, this small an error is not expected to carry over into regions far away from the beta-stable valley.

In the plot of the N–Z plane shown in Figure 2, the shaded region is the *r*-process path resulting from equilibrium calculations based on $T_9 = 1$ and $n_n = 10^{24}$ cm^{-3}. Diagonal lines indicate beta decay paths which run along isobars toward, the valley of beta stability.

In order to calculate the dynamics of the beta decay chains we have computed decay rates based on the following set of assumptions: Between the nuclei (Z, A) and $(Z + 1, A)$ every occupied state of (Z, A) can decay to about every third state of the daughter $(Z + 1, A)$ with a value of $\log ft$ which is a function of Z, ranging from 5 at $Z = 20$ to 6.5 at $Z = 100$. Since the decay energy to the ground state is typically about 15 MeV, where as the excitation in the parent nucleus is only 100 keV the decay probability from all excited states is essentially the same, and it is sufficient to consider all parent nuclei to be in the ground state. The usual approximation for the Fermi function goes as $F(Z, W) \sim W^5$ for low Z, but is low by a factor of about ten for $Z = 80$. Since our assumed ft varies in just the opposite way, errors cancel if we choose $\log ft \equiv 5$ and $F(Z, W) = W^5$. If W_0 is the nuclear mass difference and Δ is the average level spacing of the daughter nucleus, then integrating over excitation of the daughter we find

$$\lambda_\beta (Z, A) = \frac{10^{-5}}{18 \ln 2} \frac{W_0^6}{\Delta} \text{ sec}^{-1}$$

with W_0 and Δ in MeV.

Having found the decay probabilities for each isotope it is only necessary to take a sum, weighted by the equilibrium distributions found previously, to find the total decay probability for a given Z, i.e.,

$$\lambda_Z = \sum_A p(Z, A)\, \lambda_\beta(Z, A) \text{ sec}^{-1}$$

where

$$p(Z, A) = n(Z, A) / \sum_A n(Z, A).$$

Figure 3 is a plot vs. Z of half lives thus obtained for $T_9 = 1$ and $\log n = \log n_n = 24$.

The dynamic problem is then specified by the set of differential equations

$$\frac{d}{dt} n_Z(t) = \lambda_{Z-1} n_{Z-1}(t) - \lambda_Z n_Z(t)$$

with initial conditions $n_{26}(0) = \text{constant} = N_0$, and $n_{Z \neq 26}(0) = 0$. For times short enough so that no fissile nuclei are produced the Bateman (1910) solution is valid:

$$n_{26}(t) = N_0 e^{-\lambda_{26} t}$$

$$n_Z(t) = N_0 \sum_{i=26}^{Z} \left\{ \frac{\lambda_i}{\lambda_Z} \left[\prod_{\substack{j=26 \\ j \neq i}}^{Z} \frac{\lambda_j}{(\lambda_j - \lambda_i)} \right] e^{-\lambda_i t} \right\} \quad \text{for } Z > 26.$$

A second form of solution is valid for times long compared to the sum of all beta decay lives. In this case fission interrupts the neutron capture chain at a terminal value of Z, and provides a continuous insertion of material at two values of Z representing the fission fragments. Thus a cyclic flow is established. In this case the fission rate will control the flow of material in the r-process with the number of nuclei doubling in some characteristic time which we call the cycle time. Thus the time dependence of each n_Z is given by

$$n_Z(t) = n_Z^0 e^{\Lambda t}$$

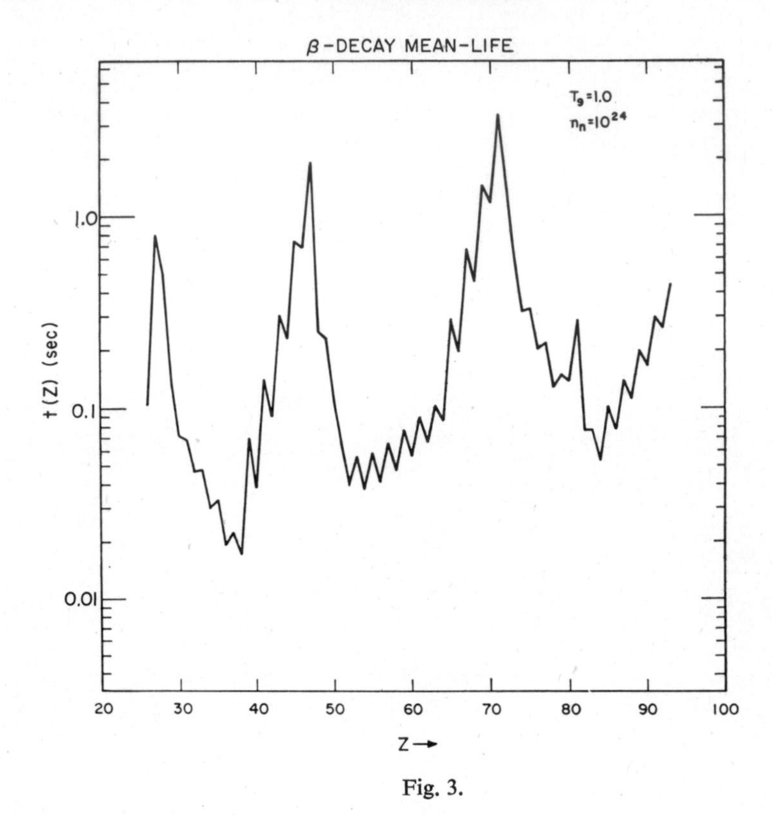

β-DECAY MEAN-LIFE
T9 = 1.0
nn = 10^24
t (Z) (sec)
1.0
0.1
0.01
20
30
40
50
60
70
80
90
100
Z →

Fig. 3.

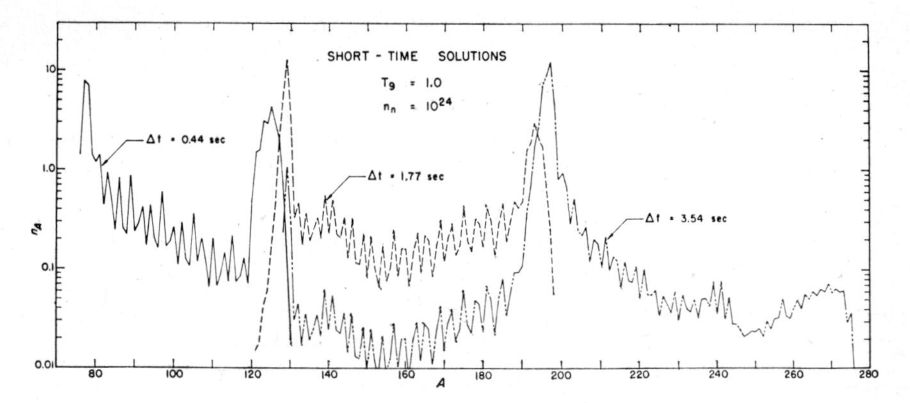

SHORT - TIME SOLUTIONS
T9 = 1.0
nn = 10^24
Δt = 0.44 sec
Δt = 1.77 sec
Δt = 3.54 sec
nA
10
1.0
0.1
0.01
80
100
120
140
160
180
200
220
240
260
280
A

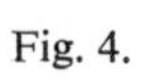

Fig. 4.

which reduces the system of differential equations to the algebraic system:

$$\Lambda\, n_Z^0 = \lambda_{Z-1} n_{Z-1}^0 - \lambda_Z n_Z^0 .$$

The cycle time is then (ln 2)/Λ. Setting an upper limit on what value of Z may be attained, the system can be solved for both Λ and the number densities. Once solutions have been found for either the long-time or short-time equations the abundance for a given A is a sum over isobars:

$$n_A = \sum_Z n_Z(t)\, p(Z, A)$$

Sample solutions are shown in Figures 4 and 5. The short-time solutions in Figure 4 are for three durations of time and illustrate the progressive evolution of the abundance peaks towards higher values of A. In this case,

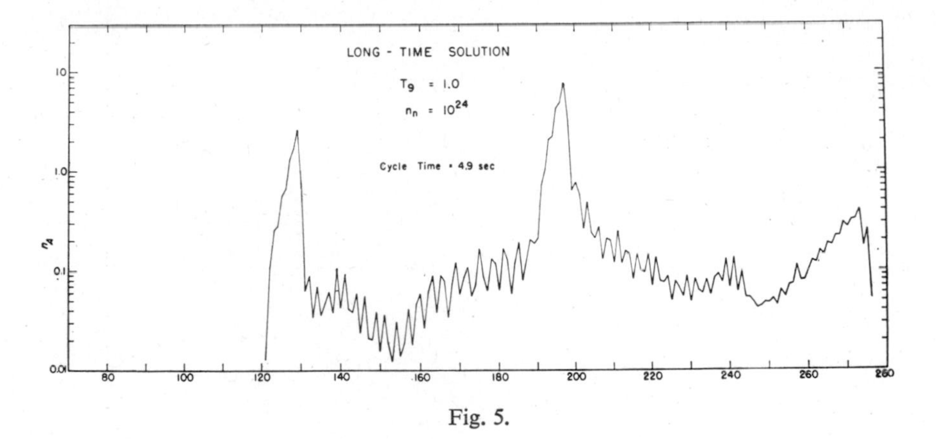

Fig. 5.

when the elapsed time exceeds 3.54 seconds fission begins and the solution merges with the long time solution of Figure 5, which has a cycle time of 4.9 seconds. On examination of a number of cases with various combination of neutron density ranging from 10^{18} cm^{-3} to 10^{32} cm^{-3} and temperatures between 0.6×10^9 °K and 3.6×10^9 °K it was found that whenever two solutions have the same cycle time then all other features are essentially identical. Thus, a given final condition can be reached from many different combinations of environment providing they have the same cycle time. For convenience in the discussion to follow, cycle times as functions of neutron density and temperature are shown in Figure 6.

In Figure 7 is shown a summary of the results obtained. The region labeled first peak indicates the range of cycle time and elapsed time Δt in which the position of the first abundance peak at $A = 80$ is reproduced along with the gentle slope between $A = 95$ and 120. A further condition which must be met is that the height of the second peak relative to the first must not be more than the observed ratio. A typical solution is shown in Figure 8 where points are the observed *r*-process abundances. Since the first peak contains

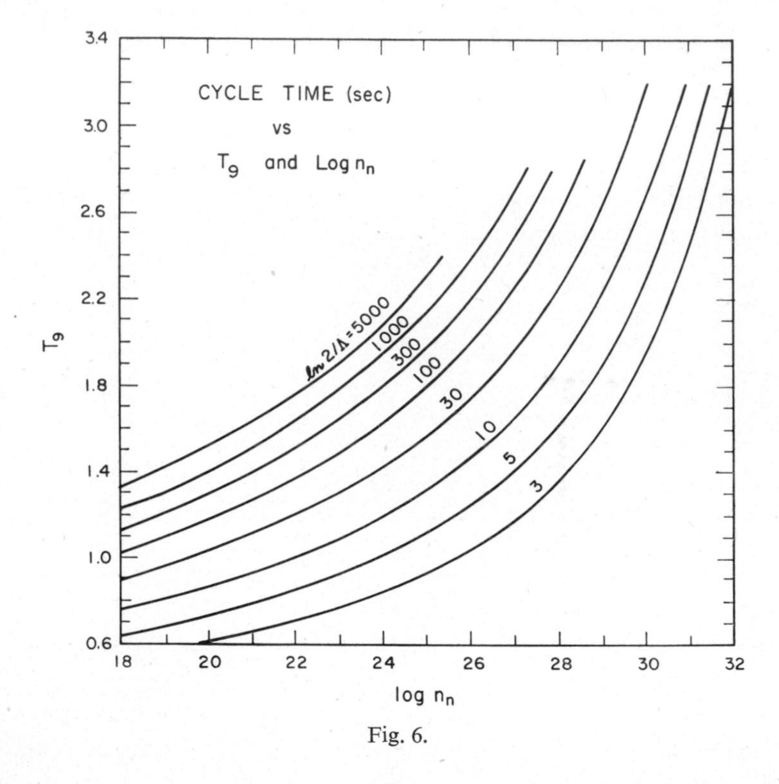

Fig. 6.

the bulk of *r*-process material it follows from Figure 7 that the majority of the material was exposed to the neutron flux for only one to five seconds, otherwise the first peak would have been destroyed.

Under none of the conditions considered did the calculation reproduce a peak near $A = 85$, nor does it seem likely that any physically reasonable mass law will do so. We therefore tentatively ascribe the krypton peak to the *s*-process until new measurements of the krypton neutron capture cross section become available.

The position of the second peak centered about $A = 130$ was reproduced for a wide range of cycle times, and hence temperature and densities. Thus, it imposed only a slight restriction on permissible environments. To produce

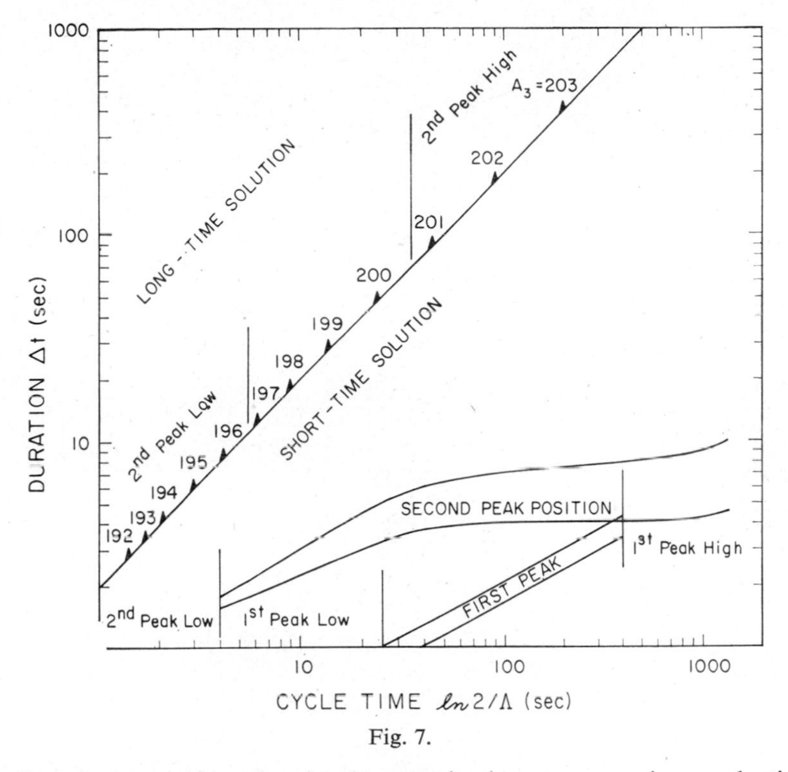

Fig. 7.

the first and second peaks simultaneously, however, requires cycle times between 350 and 400 seconds and durations of about 4 seconds. (Figure 8 is such a solution).

On the other hand, a long time solution with cycle times between 5.5 and 35 seconds can reproduce the second peak position (see Figures 7 and 5), but in such cases the third peak height is too high compared to the second. We conclude that it is possible for some second peak material to have been produced by recycling, but this remains uncertain.

The situation regarding the third peak is very uncertain. It appears that with the mass law used very short cycle times (i.e., either very high neutron densities or low temperatures) are required to get the position at $A = 195$. Because the location and height of the peak are very strong functions of the mass law, and because the r-process path passes through regions in which there are virtually no known masses, the true picture remains obscured. It is certain that that in the r-process as described here it is impossible to have material reach the third peak while any remains in the first. Thus we must say that solar r-process material was produced in two distinct types of events: one type lasting about 4 seconds to produce the first peak and most of the second and another type lasting for long enough times that cycling occurs. The location of the third peak as a function of cycle time is indicated along the line in Figure 7 which delimits the region in which the long-time solutions hold. It is seen that for cycle times of about 3 seconds the peak falls in the correct position.

We now wish to see what physical situations are needed to provide the conditions described above for the *r*-process. For synthesis of the first and second peaks a Δt of about 4 seconds is required. We shall follow Hoyle and Fowler (1963) who suggest that the site of the *r*-process is associated with the explosive outbursts of massive stars ($M > 10^4$ M$_\odot$) such as quasi-stellar objects. Since free explosion is the reverse of free fall, the time scales are comparable. From Fowler and Hoyle (1964) we find

$$\Delta t \sim 10^{14}/n_n^{1/2}.$$

For $\Delta t \sim 4$ seconds, $n_n = 5 \times 10^{26}$ cm^{-3}.

From Figure 7 a cycle time of about 350–400 seconds is required. From Figure 6 we thus obtain a temperature of 2.4×10^9 °K. Using the expressions of Fowler and Hoyle (1964) the range of stellar masses for the terminal stages of explosion which satisfy the above is

$$M \sim 10^{5\pm1}\ \mathrm{M}_\odot,$$

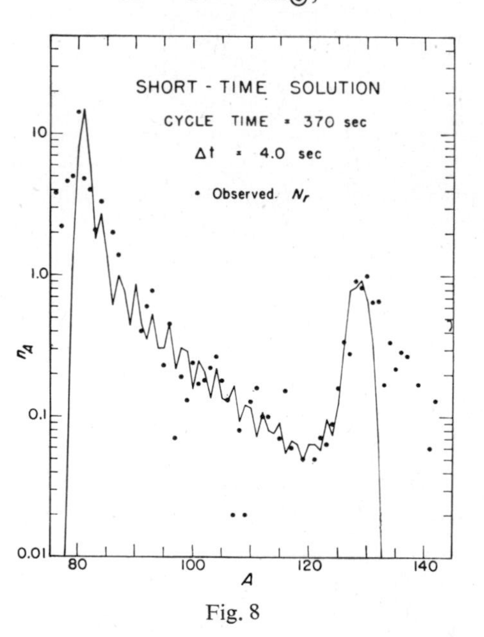

Fig. 8

which is in substantial agreement with the results of Hoyle and Fowler (1963).

If the third peak is synthesized in the same object or an object of the same mass, then a temperature of 10^9 °K and $\log n_n = 25.5$ is required in order to have a cycle time of 3 seconds.

REFERENCES

Bateman, H. (1910). *Proc. Cambridge Phil. Soc.* **15**, 423.

Becker, R. A. and Fowler, W. A. (1959). *Phys. Rev.* **115**, 1410.

Burbidge, E. M., Burbidge, G. R., Fowler, W. A. and Hoyle, F. (1957). *Rev. Mod. Phys.* **29**, 547. Called B^2FH.

Clayton, D. D., Fowler, W. A., Hull, T. E. and Zimmerman, B. A. (1961). *Ann. Phys.* **12**, 331.

Clayton, D. D. (1964). *Astrophys. J.* **139**, 637.

Fowler, W. A. and Hoyle, F. (1964). *Astrophys. J. Suppl.*, **91**, 1.

Gibbons, J. H. and Macklin, R. L. (1965). *Rev. Mod. Phys.*, **37**, 166.

Hoyle, F. and Fowler, W. A. (1963). *Nature*, **197**, 533.

Kümmel, H., Mattauch, J. H. E., Thiele, W. and Wapstra, A. H., (Vienna: Springer-Verlag, 1964). *Proceedings of the Second International Conference on Nuclidic Masses.*

Seeger, P. A., Fowler, W. A. and Clayton, D. D. (1965). *Astrophys. J. Suppl.* **97**.

Urey, H. C. (1964). *Rev. Geophys.* **2**, 1.

Recent Experiments on Multiple Neutron Capture

G. I. Bell

The formation of heavy nuclei by multiple neutron capture can take place in the intense neutron fluxes produced in thermonuclear explosions. A study of the resulting yield curves can give information concerning the neutron capture and fission cross sections for these nuclei, relevant to discussions of rapid neutron capture in stellar interiors. The Mike fusion explosion in 1952 produced nuclei with masses through $A = 255$, leading to the discovery of the elements Einsteinium and Fermium (Ghiorso et al., 1955). More recently, as part of the Atomic Energy Commission's Plowshare Program, some low-yield devices have been exploded underground for the purpose of producing heavy nuclei. In two recent experiments of this nature, Par and Barbel, heavy nuclei up to $A = 257$ were formed by the exposure of U^{238} to intense neutron fluxes.

Attempts to interpret the Mike results were complicated by the fact that the neutron fluxes varied considerably throughout the target. For the Par and Barbel devices, small targets of U^{238} were deliberately used in order to assure that the neutron flux be nearly constant throughout the sample.

The yield curves for these two devices are shown in Figure 1 (Dorn and Hoff, 1965; and Los Alamos Radiochemistry Group, 1965). The essential features of these two curves are quite similar; the yields of both even and odd mass numbers are an exponentially decreasing function of the mass number. Furthermore, the yields for odd-A nuclei are suppressed by approximately a factor of two through $A \simeq 250$, beyond which this odd-even effect is reversed.

The general character of these curves, the constancy of successive product abundance ratios, suggests that the neutron capture cross sections cannot fall off rapidly with increasing mass number. Such a conclusion was first arrived at by Cameron (1959) from a study of the Mike results, leading to a re-evaluation of the mass formula (Cameron and Elkin, 1965). The arguments leading to this conclusion are straightforward. Consider the multiple

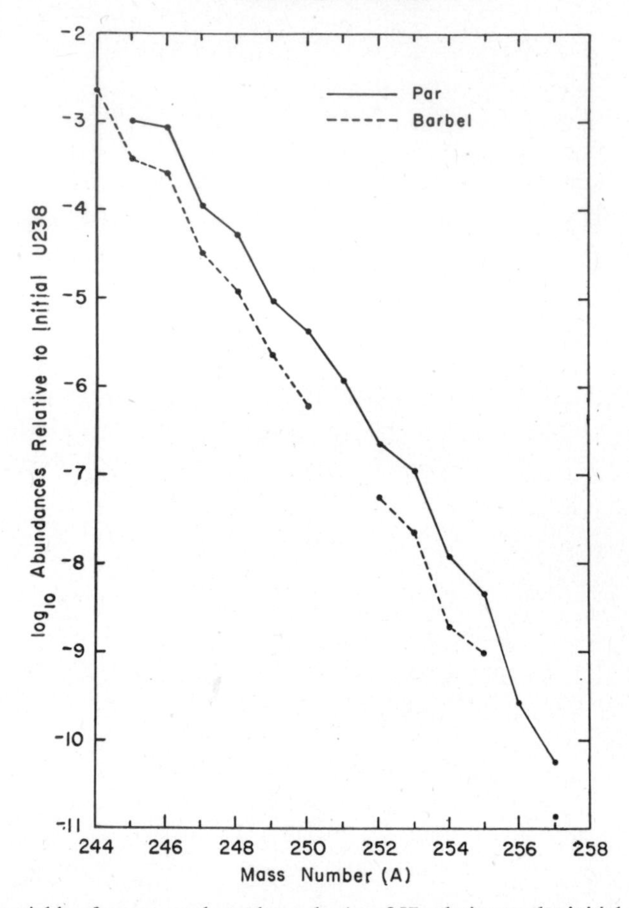

Fig. 1. The yields of mass numbers through $A = 257$ relative to the initial abundance of U^{238} are plotted as a function of mass number for the Par and Barbel devices.

capture of neutrons on a U^{238} target. Let $n(E, t)$ be the neutron density as a function of energy and time, $\sigma^{i}_{n,}\gamma(E)$ the neutron capture cross section for a uranium nucleus of mass number $238 + i$ and a neutron of energy E, and N_i the final abundance of nuclei of mass $238 + i$. If we assume that the neutron density is uniform throughout the target and that the cross sections are independent of mass number, then it follows that

$$\frac{N_i}{N_0} = \left(\frac{x^i}{i!}\right)e^{-x} \tag{1}$$

where x is a constant given by

$$\int dt \int dE \sigma_{n,\gamma}(E)\, vn(E, t) = \mathrm{x} \tag{2}$$

The ratio of the resulting abundances of two succesisve isotopes is then seen to be roughly constant:

$$\frac{N_i}{N_{i-1}} = \frac{x}{i} \tag{3}$$

These relations give values of N_i which are in rather good agreement with the experimental N_i as seen on Mike, Par, and Barbel, hence one is tempted to conclude that the cross sections are indeed constant as a function of mass number.

There is, however, another possible explanation for the relatively high yields of the heavy mass region, to which the reversal of the odd-even effect at $A \sim 250$ provides are markable clue. Diamond and Fields (1965) appear to have been the first to suggest that this reversal could be understood if the heavy mass nuclei had been produced by neutron capture in an odd-Z mass chain. The cross sections for neutron capture on odd-Z targets are generally higher than those for even-Z, due both to their higher spin values and to their lower pairing energies. Thus even a small amount of odd-Z material could dominate the abundance curve after many neutron captures. This explanation requires the formation of neptunium or protactinium at some point in the uranium capture chain by (d, n) or (n, p) reactions, respectively, and the subsequent capture of neutrons by these odd-Z nuclei (Bell, 1965). If this explanation is correct, then the high yields for masses $A > 250$ do not demand a constancy of the neutron capture cross sections for uranium through $A \sim 255$.

CALCULATION OF NEUTRON CAPTURE CROSS SECTIONS

In order to interpret these results, we have performed some detailed calculations of neutron capture cross sections. Statistical theories of neutron capture have been presented by Margolis (1952), Lane and Lynn (1957), Cameron (1963), and others. We have employed the theory due to Margolis, together with the correction, $S_1(\alpha)$, of Lane and Lynn which takes the fluctuations of the neutron width into account.

As we are interested in neutron capture at energies less than 100 keV, we consider only s and p partial waves and ignore the possibility of inelastic scattering. At these energies, the product σv is a slowly varying function of energy, and we can assume all the neutrons to have a single energy. The following choices for the nuclear spins have been emploeyd: even-even targets have spin zero, odd-A targets have spin 5/2, and odd-odd targets have spin 3.

We computed a capture cross section averaged over many resonances, assuming a constant radiation width, Γ_γ^J, for levels of given angular momentum J, and a Porter-Thomas (X^2 with one degree of freedom) distribution of neutron widths. The average neutron width per channel, $\overline{\Gamma}_n^{lJ}$ is given by

$$\overline{\Gamma}_n^{lJ} = \frac{D_J T_l(E)}{2\pi} \tag{4}$$

where D_J is the average level separation of compound nuclear levels of spin J and $T_l(E)$ is the penetration factor for neutrons of orbital angular momentum l, energy E. The cross section for capture on a target nucleus of zero spin is therefore

$$\sigma_{n,\lambda}(E) = \frac{\pi\lambda^2}{2}\left[T_0(E)\frac{2S_1(\alpha_{0,\frac{1}{2}})}{1+\xi_{\frac{1}{2}}T_0(E)} + T_1(E)\left\{\frac{2S_1(\alpha_{1,\frac{1}{2}})}{1+\xi_{\frac{1}{2}}T_1(E)} + \frac{4S_1(\alpha_{1,3/2})}{1+\xi_{3/2}T_1(E)}\right\}\right] \tag{5}$$

where λ is the reduced de Broglie wavelength for neutrons of energy E, the $S_1(\alpha_{l,J})$ are the correction factors for the neutron widths and $\alpha_{l,J}$ and ξ_J are given by:

$$\alpha_{l,J} = \frac{\Gamma_\gamma^J}{\overline{\Gamma}_n^{lJ}} \tag{6}$$

$$\xi_J = \frac{D_J}{2\pi\Gamma_\gamma^J} \tag{7}$$

The corresponding expression for capture on a spin 5/2 target is:

$$\sigma_{n,\gamma}(E) = \frac{\pi\lambda^2}{12}\left[[T_0(E)\left\{\frac{5S_1(\alpha_{0,2})}{1+\xi_2T_0(E)} + \frac{7S_1(\alpha_{0,3})}{1+\xi_3T_0(E)}\right\} + T_1(E)\frac{3S_1(\alpha_{1,1})}{1+\xi_1T_1(E)} + \frac{10S_1(2\alpha_{1,2})}{1+2\xi_2T_1(E)} + \frac{14S_1(2\alpha_{1,3})}{1+2\xi_3T_1(E)} + \frac{9S_1(\alpha_{1,4})}{1+\xi_4T_1(E)}\right] \tag{8}$$

The numerical results for a spin 3 target are nearly identical to those for spin 5/2, thus we have employed equation (8) for both cases.

The penetration factors employed in these calculations are those suggested by optical model calculations of Schrandt *et al.* (1957), being $T_0 = 0.07$ and $T_1 = 0.01$ for $E_n \sim 20$ keV.

The level spacings were calculated from the following expression:

$$D_J = \frac{3}{2J+1} e^{(2J+1)^2/140} (U')^2 \exp - 7.5\sqrt{U'} - 2] \tag{9}$$

where U' is in MeV and D_J is in eV. U' is the effective excitation energy, given by the actual excitation energy minus a pairing correction. We have employed the pairing corrections given by Newton (1956) and the neutron binding energies predicted by the atomic mass formula of Seeger (1961).

The nuclear radiation widths were computed, assuming electric dipole transitions only, from the expression (Cameron, 1963):

$$\Gamma_\gamma = \text{const. } D_0(U') \int_0^{U'} \frac{E^3\, dE}{D_0(U' - E)} \tag{10}$$

Here, $D_0(U')$ is given by equation (9) except that D_0 was taken to be constant for $U' \leqq 0.5$. These results were normalized to agree with experimental values of Γ_γ for heavy nuclei.

The 20 keV cross sections calculated from this procedure for the relevant isotopes ($238 \leqq A \leqq 257$) of protactinium, uranium and neptunium are given in Table I. The effective excitation energies are tabulated as well. The calculated value of the cross section for capture on U^{238}, $\sigma = 0.46$ barns, is in good agreement with the experimental value, $\sigma \simeq .50$ barns (Hughes and Schwartz, 1958). These results indicate that the cross sections for the uranium isotopes are not constant: rather they fall by approximately an order of magnitude from U^{238} to U^{256}.

INTERPRETATION OF THE YIELD CURVES

Having calculated the appropriate neutron capture cross sections, we can now attempt to fit the yield curves for these experiments. For the neutron capture chain, the rate of change of the abundance of a given isotope, N_i, can be written in the form

$$\frac{dN_i}{d\varphi} = N_{i-1}\sigma_{i-1} - N_i\sigma_i \tag{11}$$

Table I

Capture Cross Sections at 20 keV

A	Protoactinium		Uranium		Neptunium	
	U' (MeV)	$\sigma_{n,\gamma}$ (b)	U' (MeV)	$\sigma_{n,\gamma}$ (b)	U' (MeV)	$\sigma_{n,\gamma}$ (b)
238	4.91	1.44	3.92	0.46	5.74	2.25
239	4.05	0.68	4.52	1.06	4.85	1.38
240	4.66	1.20	3.65	0.34	5.45	1.98
241	3.79	0.51	4.26	0.84	4.55	1.10
242	4.42	0.97	3.38	0.24	5.16	1.70
243	3.55	0.38	4.01	0.65	4.27	0.84
244	4.20	0.79	3.14	0.17	4.89	1.41
245	3.33	0.28	3.79	0.51	4.00	0.64
246	3.99	0.64	2.91	0.12	4.64	1.18
247	3.12	0.21	3.58	0.40	3.75	0.48
248	3.81	0.52	2.70	0.086	4.42	0.97
249	2.94	0.16	3.38	0.30	3.52	0.37
250	3.64	0.42	2.51	0.062	4.20	0.79
251	2.77	0.12	3.21	0.24	3.31	0.28
252	3.48	0.35	2.33	0.044	4.00	0.64
253	2.62	0.096	3.05	0.19	3.11	0.21
254	3.35	0.29	2.17	0.032	3.82	0.52
255	2.48	0.075	2.91	0.15	2.95	0.16
256	3.22	0.24	2.03	0.024	3.66	0.44
257	2.55	0.058	2.78	0.13	2.77	0.12

where φ is the time integrated neutron flux

$$\varphi(t) = \int^{t} nv \, dt'$$

This system of equations can be solved numerically for specified initial values of the N_i, and a given value of the flux, φ.

In our calculations, we have treated both the integrated flux and the amounts of U, Pa and Np exposed to this flux as variables. The resulting best fits for the Par and Barbel experiments are shown in Figures 2 and 3. For both cases, it was found that a fit to the low mass region $A \lesssim 248$ can only be obtained using the uranium chain, target abundances $\sim$ 0.1 initial U^{238} abundance, and a value $\sim 7 \times 10^{24}$ cm^{-2} for the integrated neutron flux. This target abundance represents the fraction of the initial U^{238} which is present in the capture chain and is a reasonable value for the fraction which survives fission by fast neutrons (Bell, 1965). The high mass region, $A > 249$,

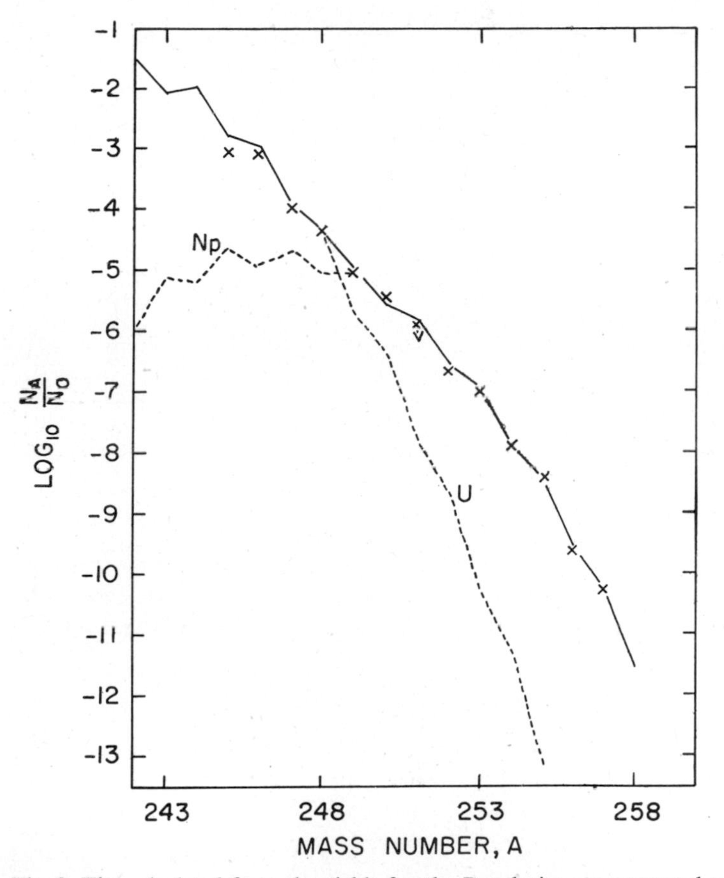

Fig. 2. The calculated fit to the yields for the Par device are compared to the experimental values (Dorn and Hoff, 1965). For this calculation the time integrated neutron flux was 7×10^{24} cm^{-2} and the amounts of uranium and neptunium exposed were 10^{-1} and 10^{-4} of the initial uranium respectively. The dashed lines show contributions of the uranium and neptunium fractions separately.

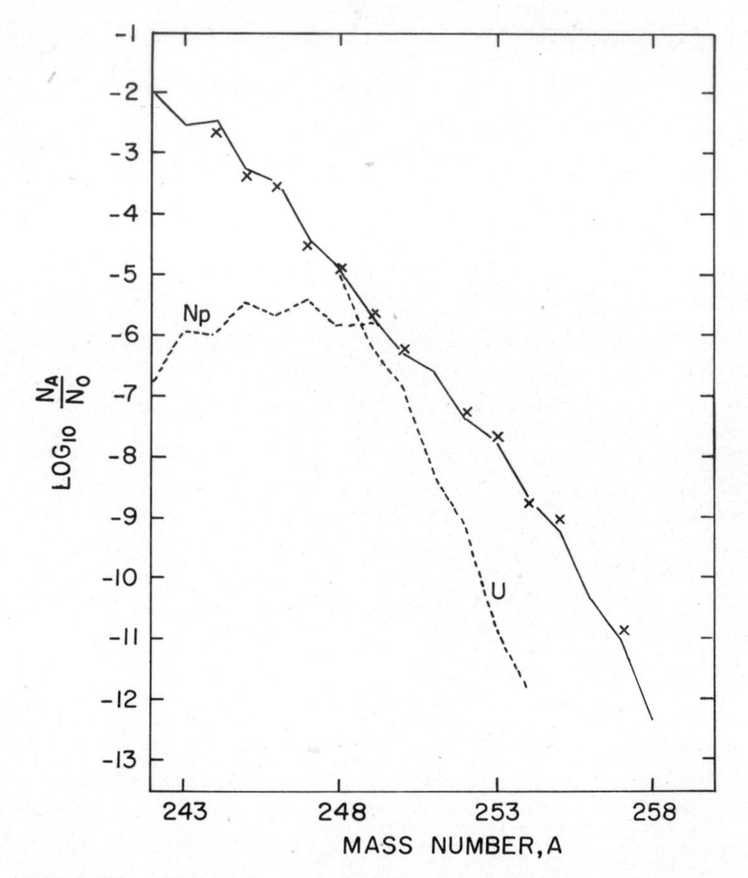

Fig. 3. The calculated fit to the yields for the Barbel device are compared to the experimental values (Los Alamos Radiochemistry Group, 1965). For this calculation the time integrated neutron flux was 7×10^{24} cm^{-2} and the amounts of uranium and neptunium exposed were 3×10^{-2} and 1.5×10^{-5} of the initial uranium, respectively.

can be fit with the same flux using the neptunium chain with Np/U $\simeq 10^{-3}$ for Par and Np/U $\simeq 5 \times 10^{-4}$ for Barbel. A fit to the high mass region by the protactinium chain would require both a larger flux, $\varphi \sim 10^{25}$ cm^{-2}, and $Pa/U \sim 10^{-2}$. The production of such large amounts of protactinium is difficult whereas the required amounts of Np may be produced mostly by (*d*, *n*) and (*d*, 2*n*) reactions (Bell, 1965).

We have also obtained a fit to the yield curve for the Mike experiment, shown in Figure 4. It was found that the mass region $245 \leqq A \leqq 249$ can

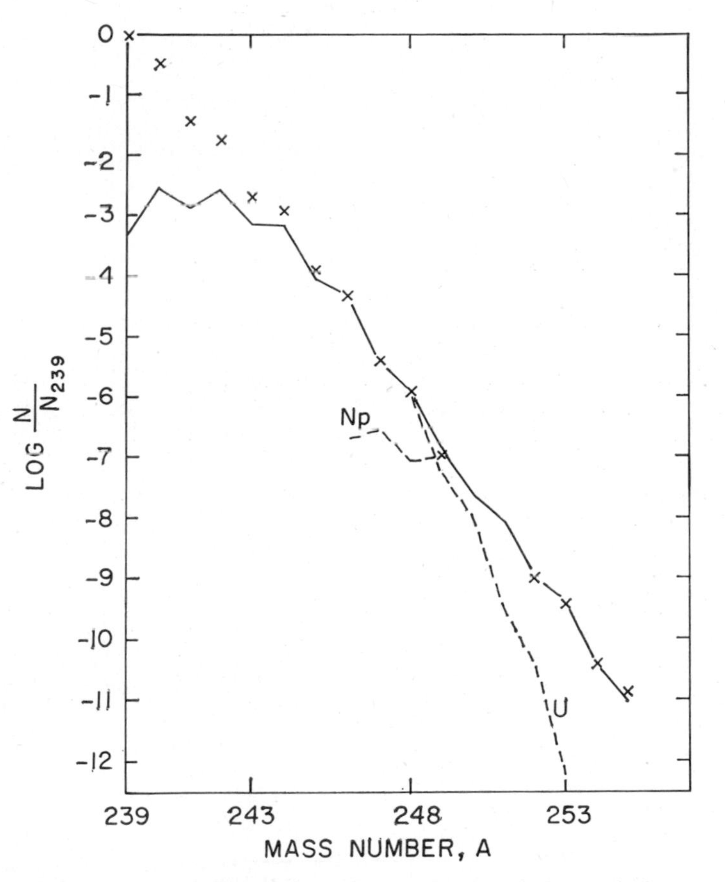

Fig. 4. The calculated fit to the yields for the Mike fusion explosion are compared to the experimental values (Diamond et al., 1960, except that the mass 255 point has been reduced by a factor 4 to take into account an unpublished reanalysis by H. Diamond). For this calculation the time integrated neutron flux was 6×10^{24} cm^{-2} and the amounts of uranium and neptunium exposed were 10^{-2} and 1.5×10^{-6} of the remaining U^{239}, respectively.

be fit by a uranium chain with $\varphi = 6 \times 10^{24}\ cm^{-2}$, and that the region $252 \leqq A \leqq 254$ can be fit by a neptunium chain, $Np/U - 1.50 \times 10^{-4}$, exposed to the same flux. The observed yields for mass $239 \leqq A \leqq 244$ can not be reproduced under these conditions. The high yields for this mass region are regarded as resulting from capture in a low flux region. This is consistent with the view that the neutron fluxes in both the Par and Barbel devices were far more uniform than that for the Mike device.

We conclude from this investigation that the yield curves for these three devices can be understood, employing conventional mass formulas, in terms of neutron capture in uranium and neptunium chains, the neptunium being formed by reactions involving energetic deuterons or tritons and uranium nuclei at some point in the uranium capture chain.

REFERENCES

Bell, G. I. (1965). *Phys. Rev.* **139**, 1207 B.

Cameron, A. G. W. (1959), *Can. J. Phys.* **37**, 322.

Cameron, A. G. W. (1963). Sec. V. M. in *Fast Neutron Physics*, edited by J. B. Marion and J. L. Fowler (Interscience, New York).

Cameron, A. G. W. and Elkin, R. M. (1965). *Can. J. Phys.* **43**, 1288.

Diamond, H., Fields, P. R., Stevens, C. S., Studier, M. H., Fried, S. M., Inghram, M. G. Hess, D. C., Pyle, G. L., Mech, J. F., Manning, W. M., Ghiorso, A., Thompson, S. G., Higgins, G. H., Seaborg, G. T., Browne, C. I., Smith, H. L. and Spence, R. W. (1960). *Phys. Rev.* **119**, 2000.

Diamond, H. and Fields, P. R. (private communication).

Dorn, D. W. and Hoff, R. W. (1965). *Phys. Rev. Letters* **14**, 440.

Ghiorso, A., Thompson, S. G., Higgins, G. H., Seaborg, G. T., Studier, M. H., Fields, P. R., Fried, S. M., Diamond, H., Mech, J. F., Pyle, G. L., Huizenga, J. R., Hirsch, A., Manning, W. M., Browne, C. I., Smith, H. L., and Spence, R. W. (1955). *Phys. Rev.* **99**, 1048.

Hughes, D. H. and Schwartz, R. B. (1958). Brockhaven Natl. Lab. Rept. BNL-325.

Lane, A. M. and Lynn, J. E. (1957). *Proc. Phys. Soc.* (*London*) A**70**, 557.

Los Alamos Radiochemistry Group (1965). *Phys. Rev. Letters* **14**, 962.

Margolis, B. (1952). *Phys. Rev.* **88**, 327.

Newton, T. D. (1956). *Can. J. Phys.* **34**, 804.

Schrandt, R. G., Beyster, J. R., Walt, M. and Salmi, E. (1957). Les Alamos Rept., LA-2099.

Seeger, P., (1961). *Nucl. Phys.* **25**, 1.

Models of Carbon Burning Stars

E. E. SALPETER

I want to report briefly on some stellar models calculated by Deinzer[1] for the "zero age carbon burning main sequence". These models represent homogeneous stars containing a fraction (by mass) x_C of C^{12} ($1 - x_C$ of O^{16}, etc.) with the $C^{12} - C^{12}$ reaction providing the only form of nuclear energy generation.

Table I below shows the central temperature T_c for various stellar masses M, for two values of the carbon abundance x_C, both for models with and without neutrino energy loss included (as a result of previous helium reactions, x_C is likely to be of order 0.5 for low masses but much smaller for high masses). As expected, values of T_c are higher, when neutrino reactions are included. The increase of T_c with decreasing x_C is also more marked with neutrinos than without.

Table I

Values of $\log_{10} T_c$ with T_c the central temperature in °K

x_c $M/M_\odot$	0.5 without	0.025 without	0.5 with	0.025 with
0.92	8.76	8.83	8.81	8.91
1.33	8.78	8.86	8.83	8.94
4.5	8.83	8.91	8.88	9.00
27	8.86	8.94	8.92	9.07

Another quantity of interest is the "minimum mass for carbon burning" M_{min}, For $M < M_{min}$ a homogeneous carbon star under gravitational contraction reach a maximum central temperature which is insufficient to burn carbon and will cool off to the white dwarf stage. We estimate that M_{min} is about 0.75 M. without neutrinos and about 0.9 $M_\odot$ with neutrinos.

[1] W. Deinzer and E. E. Salpeter, *Astrophys. J.* **142,** 813 (1965).

Summary and Impressions

G. R. BURBIDGE

To my knowledge, this is the first conference devoted entirely to the problems of nucleosynthesis. This subject is now quite a few years old. It began with the work on the hydrogen burning reactions by Bethe (1939) and von Weizsäcker (1937) before the war. The milestones since the war include, first of all, the remarkable paper by Hoyle (1946) in which he first discussed the equilibrium process and the iron-to-helium phase transition, and really laid the groundwork for much of the subsequent work in this field. This was followed by the investigation of the helium burning reactions by Salpeter (1957). Finally, over the last ten years there have been a number of papers on this subject by Cameron (1959a, b; 1963) and by Burbidge et al. (1957).

It is interesting to note the general breakdown of the twenty-three papers given at this conference. There was one, by Dr. Suess, in which it was argued that we are all wrong. There were four papers concerned with the observational data. The remaining eighteen papers were concerned with the theory, both from the point of view of stellar evolution and from the point of view of theoretical and experimental physics. It is interesting to note, as well, that the majority of these later papers are concerned with the last few months, days, seconds of the stars lifetime —a rather disproportionate amount of attention relative to the overall timescale. This is probably necessary.

I will try to give a general account of what has been said. I will begin with a rather brief comment on Dr. Suess' paper, as I find this exceedingly difficult. I simply do not feel that the difficulties he sees are really there. We have argued these points a number of times, and I do not feel I should elaborate in this discussion.

Let me turn, then, to the papers given concerning the observations. Dr. Wallerstein gave what I felt was an excellent summary of the current observational situation as it relates to stellar evolution. There are, in general, two types of evidence which must be considered in relating the theories of nucleosynthesis in stars to the observations. First of all, we believe that all of the elements heavier than hydrogen were formed in stars. We believe that the heavy element content of the galaxy has increased with time, due to these nuclear transmutations. This general correlation of heavy element content with the age of a star is well confirmed observationally. Although this type of evidence has not received much discussion, I feel it largely substantiates the type of theories which have been developed.

The presence in stars of anomalous abundances of elements which we think are largely produced by a specific nuclear burning process provides somewhat more direct evidence for nucleosynthesis in stars. Dr. Wallersteins paper was concerned largely with this type of evidence. We certainly know that both hydrogen-burning and helium-burning are currently taking place in stars. I think we also have fairly strong evidence that the *s*-process, which has been so much discussed at this conference, is taking place in the red-giant phase of stellar evolution. The presence of technetium in these stars provides a rather direct piece of evidence for the occurrence of this process.

With regard to the equilibrium process, we must still rely on the solar system abundances, and attempt to infer the conditions under which these elements were formed. With rare exceptions, we cannot measure isotope ratios in stars.

There is also the interesting question of whether *r*-process synthesis can take place in supernovae. If it can be shown that the light curves for Type I Supernovae are due in large measure, to radioactivities, then we have rather direct evidence that elements can be formed by neutron capture past the lead region right out to mass number $A \sim 260$. This hypothesis is currently somewhat uncertain due, in part, to the great variety of supernovae light curves which have been observed. It is clear that there will be energy sources other than that due to radioactivities like Cf^{254}, and detailed calculations must be carried out to determine the exact features of these light curves.

As for the observational evidence for the other processes, there is no evidence concerning the *p*-process. The process (or processes) which is responsible for the production of the light elements, lithium, beryllium, and boron, is not established, although this subject has received a great deal of attention in recent years. I think Dr. Wallerstein summarized the observational situation quite well in this regard. There is an important point concerning the lithium abundance in the sun—the new result from Kitt Peak that perhaps the feature which Greenstein and Richardson (1951) attributed to lithium is not due to lithium. This would set a limit on the lithium abundance of the sun which is somewhat lower than what was thought to be the observed abundance.

We come now to the papers of Sargent, Bidelman, and Conti, which I will discuss briefly altogether. These papers were all concerned with the anomalous abundances in stars, in particular in the peculiar *A* and B stars. Fowler et al. (1965) have recently proposed that these stars are, in fact, evolved stars, and that both interior nucleosynthesis (including a modified *r*-process) and a vast amount of surface spallation has occurred. We were led to this conclusion by studies of the anomalous abundances of these stars, as outlined by Dr. Sargent, i.e.; the overabundance of the heavy elements, particularly in the rare earth region, accompanied by a normal abundance of barium; the abundance anomalies in the iron peak region; the overabundance of silicon; the overabundance of the strontium-yttrium-zirconium peak, and the underabundance of the light elements. It should be noted that evolutionary models for these stars are not well understood. Dr. Faulkner is currently working quite extensively on these problems.

We turn now to the theoretical and experimental papers. Dr. Reeves presented the results of current research at Orsay concerning the spallation products from protons on carbon. These are important results with regard to the production of Li, Be and B in the early history of the solar system, if indeed that is when they were produced.

This was followed by three papers on stellar evolution. Dr. Iben gave a very nice account of the early stages of stellar evolution. The evolutionary state of peculiar A stars was considered by Dr. Faulkner. Finally, Dr. Stothers discussed the evolutionary characteristics of very massive stars. This is a most important subject. He whetted my appetite by talking about the Wolf-Rayet stars, but then he didn't make it quite clear where they fitted into this picture.

The Wolf-Rayet stars present a very important challenge to theories of stellar evolution and nucleosynthesis Apart from the problems associated with their positions in the H-R diagram, there are the questions regarding their peculiar composition and the extremely broad spectral lines. Limber (1964) has recently suggested that these stars are fast rotators in the late states of evolution. I do not feel competent to discuss the true compositions, but it is clear that a great deal of work is needed on the subject.

The remainder of the papers were concerned with two general topics: 1) the approach to the supernova condition and the supernova implosion and possible explosion, and 2) the detailed nuclear physics of the *s*-process.

Dr. Cameron presented a very detailed exposition on the theoretical approach to the neutron capture process, which I think is exceedingly important at this time. This work, taken together with the reported measurements of neutron capture cross sections is important to our understanding of both *s*-process and *r*-process synthesis.

I was interested in Miss Tsuruta's calculations of the equilibrium conditions and of the conditions at high densities. I thought those were very nice results, very nicely presented: presumably you are out to find neutron stars, or burst. In this regard, it is interesting to note that the question of neutron stars really has not arisen at this meeting.

We come then to the paper by Dr. Colgate and Dr. Fowler. This is, in fact the only way to approach the problem of supernovae—to follow the hydrodynamics in detail. This is a rather formidable task and, until quite recently, Dr. Colgate and his group are the only people who have been working in this field. Dr. Arnett also discussed this subject. I was intrigued by his statement that he could possibly get stars to implode, become neutrino supernovae, and then quietly go out. I am not sure this is serious possibility, but it is a very interesting one.

I think it is important to emphasize, particularly for the nuclear physicists here, what a limited amount of information we have concerning supernovae. We have a number of light curves collected over the years largely by Zwicky, and earlier by Baade. Currently there is a program in operation by which supernovae are discovered. These are normally found, of course, on other galaxies; there hasn't been a supernova in our own galaxy for 350 years. While all these light curves are for supernovae in other galaxies, the only remnants we observe are in our own galaxy. If the supernovae characteristics are extremely composition-sensitive, mass-sensitive or sensitive to the rate of star formation, we may be mistaken in our assumption that the light curves and the remnants can be correlated. We generally assume that the compositions of other galaxies are roughly comparable to our own, largely because this is the simplest possible hypothesis.

Difficulties of this kind can always arise, particularly where the observational evidence is so limited. For example, it is clear that the Crab Nebula might be abnormal in any of a number of ways. In fact, neither of the other supernova remnants we see in our galaxy of comparable age (Tycho's Supernova, 1572; Keplers Supernova, 1604) has a visible remnant. The question then arises whether there is a fundamental difference between the Crab and the other remnants or, rather, whether the Crab exploded in a region of large gas density. Did most of the material we see in the Crab result from the supernova, or is it interstellar gas excited by the explosion? In fact, we cannot at present answer these questions. This all contributes to making this a very active and a very exciting subject.

This brings me back, finally, to Dr. Suess' objections. I would like to emphasize that you cannot really say: "Perhaps it is true that nucleosynthesis takes place in stars but, as far as the solar system abundances are concerned, I can raise fundamental objections." If you do this, you raise objections to the whole approach, because most of the evidence we have concerning the abundance distribution, and all of the isotopic abundance ratios, result from studies of the solar system abundances.

REFERENCES

Bethe, H. A. (1939). *Phys. Rev.* **55**, 434.

Burbidge, E. M., Burbidge, G. R., Fowler, W. A. and Hoyle, F. (1957). *Rev. Mod. Phys.* **29**, 547.

Cameron, A. G. W. (1959a). *Astrophys. J.* **130**, 429.

Cameron, A. G. W. (1959b). *Astrophys. J.* **130**, 895.

Cameron, A. G. W. (1963). *Nuclear Astrophysics*, Yale University Lecture notes.

Fowler, W. A., Burbidge, E. M., Burbidge, G. R. and Hoyle, F. (1965). *Astrophys. J.* **142**, 423.

Greenstein, J. L. and Richardson, R. S. (1951). *Astrophys. J.* **113**, 536.

Hoyle, F. (1946). Monthly Notices Roy. Astron. Soc. **106**, 343.

Hoyle, F. (1965). *Astrophys. J.* **142**, 423.

Limber, D. N. (1964). *Astrophys. J.* **139**, 1251.

Salpeter, E. E. (1957). *Phys. Rev.* **107**, 516.

von Weizsäcker (1937). *Z. Physik* **38**, 176.

Index